For David, Tom, and Natasha

Do we recognize the dishonesty of our tradesmen with their advertisements, their pretended credit, their adulterations and false cheapness? . . .

It is not of swindlers and liars that we have need to lie in fear, but of the fact that swindling and lying are gradually becoming not abhorrent to our minds.
                        —Anthony Trollope, *The New Zealander*, 1856

Contents

Swindled

Swindled

THE DARK HISTORY OF FOOD FRAUD,
FROM POISONED CANDY TO COUNTERFEIT COFFEE

Bee Wilson

PRINCETON UNIVERSITY PRESS • PRINCETON AND OXFORD

Requests for permission to reproduce material from this work should be sent to Permissions, Princeton University Press

Published by Princeton University Press, 41 William Street, Princeton, New Jersey 08540

Library of Congress Cataloging-in-Publication Data
Wilson, Bee.
    Swindled : the dark history of food fraud, from poisoned candy to counterfeit coffee / Bee Wilson.
        p.   cm.
    Previously published: London : John Murray, 2008.
    Includes bibliographical references and index.
    ISBN 978-0-691-13820-6 (cloth : alk. paper)   1. Food contamination—History.
    2. Food industry and trade—History.   I. Title.
    TX531.W688   2008
    363.19'26—dc22   2008009688

British Library Cataloging-in-Publication Data is available

This book has been composed in Minion Pro

Printed on acid-free paper. ∞

press.princeton.edu

Printed in the United States of America

10   9   8   7   6   5   4   3   2   1

Preface

None of us likes being swindled, particularly when all we were trying to do was buy something nice to eat. The feeling—of mortification mingled with fury—is wholly disagreeable, but it is also very familiar. Being cheated over food is one of the universal human experiences. We have all been overcharged for a quart of milk; or shortchanged on a pound of strawberries; or sold an additive-laden loaf of bread that pretended to be "natural"; or served a "home-made" soup in a diner that came from a can; or eaten bacon that turned to water in the pan. Sometimes we gaze, bitterly, at the shoddy or overpriced food on our plate and wonder if there were older, simpler times, when honesty reigned and both food and its sellers had real integrity. Having researched the question for this book, I can only say that these idylls, if they existed at all, were very infrequent and short-lived. Food fraud has a long history, and it is mixed up with all the other forces—scientific, economic, political—that went together, for better and worse, to create the world we live in. In many ways, the history of food fraud is the history of the modern world.

Food has always had the power to kill as well as cure. "All things are poisons; nothing is without poison; only the dose permits something not to be poisonous," said the alchemist Paracelsus in the sixteenth century. Perhaps; but some foods are much more poisonous than others. If you drink enough carrot juice, you can make yourself ill; in 1974, a health nut called Basil Brown drank ten gallons of carrot juice in conjunction with ten thousand times the recommended dose of vitamin A; he died as a result.[1] But the really dangerous poisons are those whose dose is a normal portion. There are plenty of these in this story: children's candies dyed with copper and mercury; diseased meats dosed with chemicals to look fresh; wine sweetened with lead.

Not all poisons in food are swindles: some are entirely accidental. But what makes adulteration (deliberately tampering with food) worse than contamination is the element of intent. Behind every poisonous swindle is another human being (often, a whole team of them), who was prepared to damage your health if it meant a quick buck for them.

Adulteration is an ungainly word, and it can seem hard to pin down at times. What counts as tampering? "Am I adulterating a cake recipe when I add some extra vanilla to it?" someone asked me recently. "No, you are cooking," I replied. It is true, though, that ideas of adulteration have changed radically over time. In the 1850s, salt was listed as an "adulterant" of butter (partly because it was used to disguise butter that was going rancid). Now, salty butter is sold to those who enjoy it without any intention to deceive. By the same token, hops are now considered an essential ingredient in beer. But when first introduced, hops were viewed with deep suspicion as a rogue element in a true Englishman's ale. It took a century for hops to cease being an adulterant and become an innocent ingredient. The opposite pattern prevails now, as many ingredients once seen as harmless—saccharine, food colourings, trans fats (the hydrogenated fats that, until recently, were a routine ingredient in biscuits, cakes, and breakfast cereals)—come to be redefined as adulterants.

Yet for all these fluctuations, adulteration can be reduced to two very simple principles: poisoning and cheating. In the old common law, it was an offence to sell food that was "*not wholesome for man's body*," assuming that the seller knew what they were doing. Common law also made it an offence to sell food as something *other than what it actually was*—whether because it was of short weight, or padded or diluted, or because a cheap food had been substituted for an expensive one.[2] Although the laws against adulteration have varied widely at different times and places, they have always had these two basic ideas at their core: Thou shalt not poison and Thou shalt not cheat.

These are not just personal matters, between a buyer and a seller; they affect an entire society. Adulteration is one of the longest-standing concerns of law and government, not only for its own sake but because it impinges on so many other matters of grave importance. It is a question of public health, but also of economics. From the earliest times, governments have seen swindling as a threat to economic order and to their own authority. In addition to the more immediate victims of the crime, food cheats damage the exchequer, since they avoid paying the full taxes they would have been liable to pay on the real food or drink. Adulteration is a threat to civilized politics, too. Governments have sought to police food fraud because to permit it to carry on unpunished is a sign of anarchy. A society in which swindling is rife is one in which fundamental trust between citizens has broken down. It is therefore a vital concern of politics to stop it.

Even so, for the past two hundred years many governments have allowed swindlers to get away with outrageous crimes. This is a story of turpitude and greed, of the vile indifference with which some human beings will treat the health of others if it means making money. But it is also a story of a failure of politics; of the deep reluctance of postindustrial governments to interfere with the markets in food and drink—something earlier governments were happy to do—even when those markets have become dishonest and dangerous. The heroes of the story are not for the most part politicians, but scientists—detectives of the kitchen—who have dared to use every tool at their disposal to expose the manifold ways in which food is tampered with, padded, dyed, faked, diluted, substituted, poisoned, mislabelled, misnamed, and otherwise falsified.

Before the story begins, I should say two things about its scope as presented here. The first is that I do not write about drugs, only about food and drink, even though almost all food legislation on adulteration has also dealt with the adulteration of drugs. We are all familiar with current stories of the hair-raising trade in pirated and fake drugs, much of it in the developing world. We also know about the perils of the black market in illegal drugs in our own society. Like food adulteration, drug adulteration has a long history. Much of the traffic in falsified drugs parallels the commerce of fake food; the adulteration of illegal substances has echoes of earlier adulterations of alcoholic beverages. But to do justice to the history would take a whole other book (which has already been written, admirably, by Ernst Stieb, in 1966: *Drug Adulteration*).

One other thing: it will be seen that I write a disproportionate amount about the faked food of Britain and America. This is not just because I am British, and therefore more intimately acquainted with the bad food of these islands than of anywhere else. There is a historical reason, too. Adulteration on an endemic scale is a disease of industrialized cities, coupled with a relatively noninterventionist state. Britain was the first to acquire these two conditions at the same time, which goes some way toward explaining why we British have—over the past two centuries—endured a more debased diet than other nations in Europe. The United States soon followed suit— with the horrors of the New York swill milk scandal in the 1850s and the gruesome jungle of Upton Sinclair's Packingtown in the early 1900s. This pattern of early endemic adulteration explains something of America's predicament with food, up to the present day.

For these reasons, the book starts in Britain, in 1820, with a German scientist who had the vision and courage to point out just how bad the swindling had become.

Swindled

1

GERMAN HAM AND ENGLISH PICKLES

With Bentham bewilder, with Buonaparte frighten,
With Accum astonish . . . —James Smith, *Milk and Honey* (1840)

In the history of food adulteration, there are two stages: before 1820 and after 1820; before Accum and after Accum. It was only after 1820 that any sort of concerted fight against poisonous or superfluous additions to food in the modern Western world began, a fact that is entirely due to the appearance that year (in both Britain and the United States) of a single small book entitled *A Treatise on Adulterations of Food, and Culinary Poisons*, written by the expatriate German chemist Frederick Accum (1769–1838). It would be an exaggeration to say that this book changed everything; after it was published, the swindlers carried on swindling, and more often than not they still got away with it; no food laws were changed on account of Accum; and Accum himself, though initially feted, later suffered a total personal disgrace. But his treatise finally opened people's eyes to the fact that almost everything sold as food and drink in modern industrial cities was not what it seemed; and by being not what it seemed, it could kill them.

A Westphalian by birth (his real name was Friedrich), though a Londoner by choice, Frederick Accum was a man who loved his food. He was robustly fond of good healthy bread (wholemeal, not white), smoked ham, aromatic black coffee, and properly made jams and conserves, simmered from peaches, cherries, pineapples, quinces, and plums, or, when ripe, from delicious orange apricots.[1] Accum's attitude toward food was not that of a French gastronome, who looks down his nose at anyone who fails to yelp with delight at the sight of a truffled partridge. His appetites were less pretentious and more Germanic than this. A malty pint of beer; a bowl of sauerkraut made from white winter cabbages and caraway seeds; a crunchy pickled cucumber seasoned with pimento; a "light, flaky pie-crust"—these were a few of his favourite things. Yet there was nothing cavalier about Accum's approach to eating. He insisted that you could be as exact and particular about boiling potatoes as you might be about dressing fancy steaks; and only snobs would pretend otherwise. For Accum, this was not simply a question of taste; it was also a question of science. Cooks, he held, were chemists, and the kitchen was a "chemical laboratory." This was something Accum was well placed to judge, since, at the height of his career, in 1820, he himself was perhaps the most distinguished and certainly the most famous chemist in London, at a time when chemistry was at its zenith, and chemists were true celebrities.

As both a lover of food and a chemist, Accum believed in "precision in mixing ingredients." And, as both a lover of food and a chemist, he shuddered with moral indignation at those "respectable" criminals who tampered with food for the sake of profit. Fortunately, Accum did not keep his indignation to himself but wrote it up in his treatise, showing that countless foodstuffs were routinely falsified in ways that were at best dishonest and at worst poisonous. "It would be difficult," he wrote, "to mention a single article of food which is not to be met with in an adulterated state; and there are some substances which are scarcely ever to be procured genuine."[2] The work in which these words first appeared sold a thousand copies in a month (a substantial figure for the time) and went on to sell countless thousands more.

To read the reviews of Accum's treatise is to get a sense of a sudden collective sickening at the thought of how basic foods were falsified. "Since we read [Accum's] book," wrote a reviewer in *Blackwood's Edinburgh Magazine*, "our appetite has visibly decreased . . . yesterday . . . we turned pale in the act of eating a custard."[3] Another reviewer, in the *Literary Gazette*, complained that "It is so horribly pleasant to reflect how we are in this way be-swindled, be-trayed, be-drugged and be-devilled, that we are almost angry with Mr Accum for the great service he has done the community by opening our eyes, at the risk of shutting our mouths forever." The reviewer went on to lament:

> Our pickles are made green by copper; our vinegar rendered sharp by sulphuric acid; our cream composed of rice powder or arrow root in bad milk; our comfits mixed of sugar, starch and clay, and coloured with preparations of copper and lead; our catsup often formed of the dregs of distilled vinegar with a decoction of the outer green husk of the walnuts, and seasoned with all-spice, Cayenne, pimento, onions and common salt—or, if founded on mushrooms, done with those in a putrefactive state remaining unsold at market; our mustard a compound of mustard, wheaten flour, Cayenne, bay salt, radish seed, turmeric and pease flour; and our citric acid, our lemonade, and our punch, to refresh or exhilarate, usually cheap tartareous acid modified for the occasion.[4]

This is a fair summary of Accum's book, which called for "all classes of the community to cooperate" to abolish the "nefarious traffic and deception" of adulterating food and drink.[5] Accum describes children's custards poisoned with laurel leaves, tea falsified with sloe leaves, lozenges made from pipe clay, pepper mixed with floor sweepings, pickles coloured green with copper and sweets dyed red with lead. "Good heavens!" exclaimed one reader. "Is there no end to these infamous doings? Does nothing pure or unpoisoned come to our tables?" In truth, there *was* almost no end to the scandals Accum uncovered. His book was greeted with shock and consternation. It has been said that no chemistry book was ever so widely discussed.

This shocked reaction was exactly what Accum had sought. "THERE IS DEATH IN THE POT," read a motto in large letters on the side of an urn on the title page. In case anyone missed the point, the urn was draped in a shroud, with a gruesome skull set above it, and two serpents slithering around. Accum repeated the theme in his text. "We may exclaim with the sons of the prophet," he wrote, "*There is death in the pot*," reminding his readers of the line's biblical origins (in II Kings 4:40). The slogan "Death in the pot" would become a rallying cry for food safety campaigners of the nineteenth century, but it was never employed with such biting moral outrage as by Accum. "Feelings of regret and disgust" were, for him, an entirely appropriate response to adulteration. The "nefarious practice," he complained, was applied not just to "the luxuries of life" but to basic necessities, such as bread, which was commonly mixed with alum to give it a spurious appearance of whiteness. The motive behind adulteration was "the eager and insatiable thirst for gain," a greed so overwhelming that "the possible sacrifice of a fellow creature's life is a secondary consideration."[6] "It may be justly observed," Accum sorrowfully remarked, just in case any of his readers might have failed to get the message, "that 'in the midst of life we are in death.'"[7]

"Death in the Pot": a detail from the frontispiece of Accum's *Treatise on Adulterations of Food and Drink* (1822 edition).

What made Accum's *Treatise* so compelling? It was not that people had been entirely unaware of adulteration in food before 1820. He himself said, in his preface, that "every person" was aware that bread, beer, wine, and "other substances" were frequently adulterated.[8] Complaints about watered-down or doctored wine go all the way back to the ancient Romans, as the next chapter will discuss in more detail. More recently, in the eighteenth century, there had been countless rumours and satires on the contamination of food. The subject of adulteration found wonderful expression in the writing of the novelist Tobias Smollett, who described in his novel *Humphry Clinker* the foul and debased foods of London, compared to the bucolic simplicity of the country where the chickens are free, the game are fresh from the moors, and the vegetables, herbs, and salads are picked straight from the garden. As Smollett describes it, London is a place where strawberries are washed in spit, where vegetables are cooked with brass to make them green, and where milk carried in open pails through the streets is contaminated with the "spewings of infants," "spittle, snot and tobacco-quids from foot passengers," "spatterings from coach wheels, dirt and trash chucked into it by rogueish boys for the joke's sake," and even "frothed with bruised snails." The bread in London is "a deleterious paste, mixed up with chalk, alum and bone-ashes; insipid to the taste and destructive to the constitution." The wine is a "vile, unpalatable and pernicious sophistication, balderdashed with cider, corn-spirit and the juice of sloes."[9]

This is all very disgusting, but no one reading Smollett at the time would have believed that London food was really so bad. He exaggerates for the sake of comedy. What was so startling for Accum's earliest readers was the discovery that many of the adulterations that people had assumed to be comic distortions were actually true. The editor of the *Literary Gazette* commented, while reviewing Accum's *Treatise*:

> One has laughed at the whimsical description of the cheats in *Humphrey Clinker* but it is too serious for a joke to see that in almost everything which we eat or drink, we are condemned to swallow swindling, if not poison—that all the items of metropolitan, and many of country consumption, are deteriorated, deprived of nutritious properties, or rendered obnoxious to humanity, by the vile arts and merciless sophistications of their sellers.[10]

Accum's genius was to make readers see that "swallowing swindling" really was "too serious for a joke." That he managed this was due in part to the times in which he lived—times when the possibilities for adulterating food multiplied as science and industry grew—and in part to his own outstanding talents as a publicist. Accum was the perfect commentator on this new rash of skulduggery, since he was a man whose passion was the science and industry of modern Britain, who nevertheless saw that science and industry could be used to do damaging things to food. The story of this book is largely the battle between the science of deception and the science of detection. Accum's *Treatise* represents the beginning of this struggle.

Portrait of the chemist Friedrich Christian (Frederick) Accum (1769–1838), who first opened the eyes of the British public to the extent of food adulteration. An engraving from the *European Magazine* (1820) at the height of Accum's fame.

The Glorious Career of Frederick Accum

London in Accum's lifetime was a city where outsiders could forge their way to the top very quickly, if they had enough bravado and talent. In this centre of commerce and relative tolerance, Accum was one of many Germans who made their way to prominence. There had been a German Lutheran church in the East End since 1762 and a German school attached to it. Accum counted among his London friends the German publisher and inventor Rudolph Ackermann (1764–1834), famous for his lithographic prints. But he was by no means part of a Germanic clique. Accum had a knack for making contacts in every walk of British life: lawyers, scientists, politicians, aristocrats, men of letters; and charladies, as we shall see.

Bad food was not the first public cause Accum took up. A man with limitless reserves of energy, charisma, and a swaggering belief that he could make himself the master of every branch of chemistry and turn it to profit, he had already secured a place in history by overcoming public prejudice against gas lighting. It was largely thanks to Accum that by 1815 the streets of Westminster were lit by gas lamps rather than, as previously, by lanterns. Yet this was not all. Accum was also famous as a chemical lecturer and teacher, as a purveyor of chemical equipment, as a royal apothecary, as a popularizer of the ideas of Lavoisier, and as a writer on every subject from analytical mineralogy and crystallography to vanilla pods. "In the whole history of chemistry," wrote C. A. Browne in 1925, "there is no one who has attempted to discharge so many different roles as Accum."[11] Most historians no longer believe in the notion of a single "industrial revolution"; but if we might still speak of such a thing, Accum was the industrial revolution in microcosm, a man in whom knowledge and trade, enlightenment and profit, plus a fierce pride in the riches and power of Great Britain, coalesced in one single, handsome, restless figure.

His gas lighting triumph shows how Accum's talents combined to achieve spectacular results. Gas lighting had first been demonstrated in Paris in 1786 by the French innovator Philippe Lebon, but Lebon's "thermolampes" had not been put to practical use. They certainly hadn't made it to London, whose streets and houses were still lit mostly by whale oil, tallow, or beeswax. Then, in the winter of 1803–4, a German based in England called F. A. Winsor (another Friedrich like Accum who had anglicized his name to Frederick) gave a series of spectacular demonstrations on the benefits of gas lighting at the old Lyceum theatre on the Strand.[12] One witness described how the large theatre was "most brilliantly illuminated by inflammable air," with tubes of gas fixed round the ceiling, the boxes, and the stage, supplied from a reservoir below. Soon after, Winsor obtained a patent for his method of gas lighting and tried to set up a company for lighting the buildings and streets of London.

Ornate gas lights from Accum's *Practical Treatise on Gas-Light* (1815).

But there was widespread resistance to the idea of burning gas for light. Ordinary people feared the stench and that gas would escape and poison them. British seamen opposed gas lighting, fearing that the decline in demand for whale oil would put them out of a job. Perhaps most significantly, many distinguished scientists also cast doubt on whether gas illumination could ever be safe, including Sir Humphry Davy (1778–1829), the electrochemical genius who later invented the miner's safety lamp. It didn't help that Winsor was himself no scientist, but a self-aggrandizing buffoon. In 1807, he gave his proposed company the ridiculous title of the New Patriotic Imperial and National Light and Heat Company and boasted that annual profits would amount to £229 million (around £15 billion in today's money, using the retail price index, the most modest index of calculation)—a transparently self-serving and counterproductive attempt to garner investors.[13]

It was here that Accum stepped in. An accredited and well-connected scientist, he had a much better sense of what it would take to make gas lighting a going concern. Like Winsor, he scented the profit that could be made from gas, but went about it in a much more effective manner. He brushed up an earlier interest in the chemistry of gas, spending months experimenting with gas stoves, measuring gas flames against tallow flames, and distilling sticky coal tar with the consistency of treacle. He then presented himself before Parliament as an expert witness on the safety of Winsor's method of gas lighting. In 1809, Accum assured the House of Commons that, based on his extensive experiments, there was "no smell" in gas at the time of combustion, if it was properly done and "no danger" of the gas bursting the pipe; gas illumination, he insisted, was not just as safe as tallow light, it was greatly superior to it. Compared to the constant dangers of fires caused by guttering candles in enclosed spaces, gaslight was cleaner and more dependable, contained in its glass dome.[14] It was the right light for a modern society. In 1810, Parliament passed a bill permitting the incorporation of Winsor's company; the name of Frederick Accum, "practical chymist," appeared on the first board of directors. In 1813, Westminster bridge was lit by gas, and by 1815 thirty miles of gas main had been laid in London.

Accum was now in demand throughout England as a coal-gas expert. In 1815, he published his *Practical Treatise on Gas-Light*, the first work in any language on the subject, beautifully illustrated with pictures of gas chandeliers and lamps, published by Accum's great friend, Rudolph Ackermann. In this work, one can see Accum's passionate belief in industrial progress. Accum pleads with his readers to ignore the "common clamour" that rises up against all "improvements in machinery," whether the steam engine, new spinning and threshing machines, or gaslight. "It ought never to be forgotten," he wrote, "that it is to manufactories carried on by machinery, and abridgment of labour, that this country is indebted for her riches, independence, and prominent station among the nations of the world."[15] Well he might believe in progress and in Britannia, given his own personal progression from German obscurity to British fame and wealth.

Accum's career was one of dramatic upward mobility, from humble origins in Westphalia to a glittering and lucrative position as part of the scientific establishment of London. The sixth of seven children, he was born in Buckeburg on 29 March 1769—the birth year of both Napoleon and Wellington, as well as the great scientist Cuvier, and the year that "James Watt patented his steam engine and Arkwright his spinning frame."[16] But nothing suggested that Accum's birth would lead to similar greatness. Only two of his siblings reached maturity, a sister, Wilhelmina, and a brother, Phillip; the other four died, two of smallpox. Accum's father was a soapmaker, a converted Jew, born Herz Marcus, who at the age of twenty-eight changed his name to Christian Accum, probably for reasons of love; soon afterwards, he married Accum's mother, Judith Suzanne Marthe Bert la Motte, a devout Huguenot. Christian died when Accum was only three, so it was his mother, Judith, who brought him up. Not much is known of Accum's childhood, except that he attended the local gymnasium, where he would have studied Homer and Herodotus, Cicero and Tacitus, but no science. His inclination towards chemistry seems to have been nurtured more by familial influences. Watching his father, and later his older brother Phillip, making soap must have been one influence, as all Accum scholars have noted.[17] The process of saponification is a very immediate form of chemistry. Young Accum must have known his acid from his alkali, must have seen the magic of how potash could turn crude hog's lard into soap. Yet he did not choose to follow his brother into the family business. What Accum scholars don't say is that his love of chemistry was undoubtedly encouraged also by watching his mother cook.

"The inhabitants of Westphalia," he later recalled, "are a hardy and robust people, capable of enduring the greatest fatigues, [who] live on a coarse brown rye bread."[18] He probably meant pumpernickel, a Westphalian speciality famous even then outside Germany, whose strange dense texture was said to owe a lot to the characteristics of the local grain.[19] Accum himself came to prefer a slightly lighter wholemeal sour dough bread; but some of his tastes remained distinctively Westphalian. Long after he had settled in London and left Westphalia behind, Accum continued to extol the excellence of Westphalian ham, that smoky dried delicacy flavoured with juniper whose reputation then exceeded that of the hams of Bayonne and Parma. The nineteenth-century German food writer Rumohr called Westphalian ham one of those products that is "unique, incomparable, unrivalled."[20] A contemporary English cookery book said that "it cannot . . . be denied that the Westphalian hams, made from wild boar, have a richness and flavour which cannot be completely imparted to the flesh of the finest and fattest hogs."[21] Often, though, the hams were made from hogs and not boar. Accum wrote that "families [in Westphalia] that kill one or more hogs a year . . . have a closet in the garret, joining the chimney, made tight, to retain smoke, in which they hang their hams and bacon to dry; and out of the effect of the fire, they may be gradually dried by the wood smoke, and not the heat." This is surely autobiography as well as description. The Accum family would have needed to kill hogs for their lard for soap; delicious ham must have been a side effect.

As well as feeding him well, Accum's mother seems to have been the one who secured him his entrée to London life. Judith Accum was acquainted with the Brande family, who, thanks to the Hanoverian connection, were apothecaries to King George III of England. After leaving school, Accum took an apprenticeship at the Brande family pharmacy in Hanover. At this time, writes one historian of science, "the pharmacist's shop was virtually the only place to gain a practical knowledge of chemistry."[22] Young Accum must have done his job well, because in 1793, at age twenty-four, he was transferred to the London branch, in Arlington Street, just off Piccadilly and round the corner from St James's Palace. From here, he flung himself with great enthusiasm into the blossoming world of British science. He attended lectures at the School of Anatomy in Windmill Street, where he was taken under the wing of the physician Anthony Carlisle, who introduced him to other scientists, including William Nicholson, the founder of the *Journal of Natural Philosophy, Chemistry and the Arts*, where Accum published his early articles on the adulteration of drugs and on the scientific properties of vanilla pods.

11

Little by little, he made a name for himself. Just seven years after his arrival in London, Accum was setting up his own business in Old Compton Street, Soho, as a supplier of chemical apparatus. This is where he lived for the remainder of his time in England—the twenty years until the publication of the *Treatise on Adulterations*.

Accum's public life was very different from the narrow laboratory existence of today's academic chemist. He campaigned; he performed; he advertised; he showed off; he shared his joy of curious experiments involving Bunsen burners with the general public. He combined a fine mind with a popular touch. His aim, he wrote, was

Accum giving a public lecture at the Surrey Institute, Blackfriars Road, London, ca. 1810, by the satirist Thomas Rowlandson.

to mingle "chemical science with rational amusement."[23] He was said to have great personal charm, a fact confirmed by the warm defences of him issued by friends following the scandal that finally ruined his career in England. There are several illustrations of Accum giving public chemistry demonstrations at the Surrey Institution, where he had become professor of chemistry. A rapt audience looks on from gallery and stalls; society ladies lean forward eagerly over the balcony to get a better look. Were it not for Accum's Regency attire—with his Byronic dark hair, beetle brow, full lips and Romantic collar—he could almost be a modern celebrity chef doing a cookery show. He is pouring some substance from a great height into another substance, with a flourish.

Chemistry, then at the height of its fashion, suddenly seemed relevant to all aspects of daily life. As one journalist wrote in 1820, chemistry was "intimately connected with that enthusiasm and laudable desire for exploring the productions of nature, which characterize the age in which we live ... chemistry within our own times has become a central science."[24] Accum himself was seen as the "pet chemist" of London.[25] Chemistry governed the twin passions of

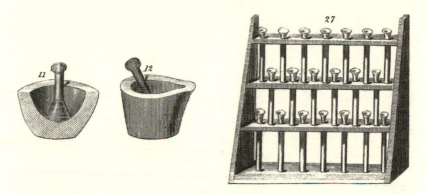

Some chemical instruments from Accum's *Explanatory Dictionary* of chemical apparatus (1824). Accum sold this kind of equipment—at a nice profit—from his premises on Old Compton Street.

Regency England, for industry and hygiene. Chemical advance powered the industrial revolution, but it was also chemistry that cleaned up the mess the factories left behind: the stinking air, rank water, and overflowing sewers. In the works of Accum, there is a fervent belief in the capacity of chemistry to make life better.

Most of his countless projects combined enlightenment and commerce. He believed in scientific truth but did not mind making money out of it. He was "obliging and kind-hearted" but also arrogant and reckless.[26] From 1800 onwards, he took on private pupils, for a fee of 160 guineas a year, a very considerable sum in those days (equivalent to around £11,000 in today's money); many of his early students went on to become distinguished scientists in the United States—Professor Benjamin Silliman of Yale, Professor William Peck of Harvard, Professor James Freeman Dana of Dartmouth. Accum was one of the first scientists to pioneer the notion of the chemistry chest, a forerunner of the child's chemistry set, although his were designed for "Philosophical gentlemen." In his book *Chemical Amusement* (1817), he wrote up instructions for "amusing" chemistry experiments that could be safely done at home. These sound marvellous, the kind of mad-scientist experiments you long to do at school but are rarely allowed to: "To melt a coin in a nut-shell"; "A fountain of fire"; "To produce an emerald green flame"; "To render bodies luminous in the dark"; "To make indelible ink for marking linen." At the back of the book is a catalogue for the apparatus and instruments needed to do the experiments, all of which can be bought—conveniently—from Accum's own shop: test tubes, hydrometers, prisms, mortars, moveable universal furnaces ranging in price from £6 16 s. to £8 8 s., as well as whole chemistry sets. He was constantly coming up with new chemical devices and materials to sell to the public. "As a manufacturer of new and improved chemical apparatus, Accum's name was for many years almost a household word . . . Accum's gasometer and pneumatic trough were in demand for fifty years."[27] If anyone should question whether science was best served by so close an alliance with trade, Accum would say, not a little pompously, that he who manufactures "an article of use to the sciences, which could not before be purchased, is a benefactor to the public."[28]

Sometimes, though, his eye for easy money took over to an extent that even Accum, for all his self-regard, might have found hard to dress up as philanthropy. One of his friends found Accum in his laboratory one day, in a state of "high glee" at a bargain he had made with Mr. Pitt, the prime minister. Pitt had come in to order a large shipment of chemical apparatus to go to Pondicherry in India. Accum was chuckling because he had played Pitt for a fool, by taking "this opportunity to sweep my garrets of all my old apparatus and odds and ends that had been accumulating for years" and to sell these oddments at an exorbitant price to the British government.[29] Interestingly, this puerile little trick did not seem morally wrong to Accum, even though he raged against the duplicity of shopkeepers; but then, he saw nothing wrong in making fun of the high and mighty, while the thought of deceiving some poor worker who only wanted to feed his family revolted him.

The Chemistry of Adulteration

Accum's world of fun, flames, and commerce was a long way from that of the Royal Academy and the other pillars of British science. Yet to a remarkable extent, Accum had managed to crack this milieu too. He may never have achieved the gravitas of a Sir Humphrey Davy, but he was a fellow of the Linnaean Society and a member of the Royal Academies of both Britain and Ireland. In 1803, he published a *System of Theoretical and Practical Chemistry*, introducing the new concepts of Lavoisier to an English audience. Accum was fortunate to have been born at a thrilling time for chemistry. As one historian of science writes, "In the space of about twenty years commencing in 1770, the science of chemistry experienced a change more complete and more fundamental than any that had occurred before or has occurred since."[30] This period saw the emergence of many of the great names in chemistry—Galvani and his twitching frog's leg, Scheele, Cavendish, Priestley, Bergman, Klaproth—but none was so great as Antoine Lavoisier, Accum's hero, who was beheaded during the French Revolution.

In 1769, the year of Accum's birth, chemistry was in many respects still alchemy. The names that chemists used for their materials were quaint and confusing—"butter of arsenic," for example, or "liver of sulphur." Central to the theory of combustion at this time was an entirely fictitious substance called phlogiston: a colourless, odourless, weightless substance believed to be present in all flammable materials. Materials containing this substance were called "phlogisticated"; when burned, they were said to be "dephlogisticated." All kinds of sensible people believed in this theory, until Lavoisier used oxygen to come up with the true theory of combustion. Lavoisier, moreover, swept aside the old alchemical names and unified chemistry into a single modern science. Lavoisier also prefigured Accum in making the first attempts to analyse the composition of alcohol, burning spirit of wine over mercury. He broke olive oil down into hydrogen and carbon. He shared Accum's interest in fruit; a memoir of 1786 by Lavoisier on the nature of organic acids analysed acids produced from pomegranates and barberries, cherries and currants, peaches, apricots, and pears.

Accum's *Treatise on Adulteration* was very much part of this newly prestigious chemistry; he saw clearly that chemical science was both the source of much adulteration and the only way of combating the growing use of adulteration. How "lamentable" it was, he thought, that chemistry, which ought to serve the "useful purposes of life" had been "perverted into an auxiliary of this nefarious traffic." Many "wholesale manufacturing chemists" occupied themselves in crystallizing alum, knowing full well that it would be used by bakers to falsify bread. On the other hand, "happily for the science," chemistry could also be "converted into a means of detecting the abuse."[31] As a later chemist wrote, analytical chemistry had the power to be "the great enemy of adulteration."[32] Before 1820, if you wanted to find out whether a certain food was pure or not, you would most likely use your eyes, nose, and tongue. If milk tasted thin and looked bluish, you might surmise it had been watered down. If coffee was too bitter, you might guess it had been tainted with chicory. If lemonade tasted too acidic and was sold too cheap, it wouldn't take a genius to figure out that it had been made with tartaric acid instead of lemons. This sort of common-sense testing of food is probably the method most employed even now to judge the quality of food. It is called the "organoleptic" approach, and so long as the food being judged is fairly natural and simple, it can work quite well. If you compared two eggs, one of inferior quality and one spanking fresh from some well-looked-after hens, you would know at once which was the good one—all your senses would tell you that the egg whose yolk was rich, orange, and flavourful was better than the anaemic one with its watery whites. But the organoleptic approach has its limits: it depends entirely on the person testing the food knowing what a certain substance *should* taste, smell, or look like; and it doesn't work so well when food has been tampered with in clever and subtle ways.

By 1820, Accum wrote, adulteration had reached "such a perfection of ingenuity" that "spurious articles of various kinds are every where to be found, made up so skilfully as to baffle the discrimination of the most skilful judges." Modern food fraudsters were using a little chemical know-how to come up with ever more cunning ways of tarting up cheap ingredients to look as good as new. If they had some old cayenne pepper on their hands, they might touch it up with red lead (just as modern fraudsters touch up stale chilli pepper with red azo dyes). Conversely, if they had some young, raw brandy that they wanted to pass off as the finest aged cognac, they might, wrote Accum, use "tincture of raisin stones" to give the brandy a "ripe taste." These new forms of adulteration made a mockery of the old organoleptic tests. Take cream, for example. By sniffing a jug of cream, you can usually tell if it is fresh; and by looking at the consistency, you can attempt to judge how rich it is—the thicker the cream, the richer and more desirable. But what if the thickness in the cream came not from butterfat but from the addition of rice flour or arrowroot, a common trick in Accum's day? Could you still trust your eyes to know the difference? With the swindlers raising their game, the process of detection had to become cleverer as well. Chemistry needed to be fought with chemistry. In the case of the adulterated cream, Accum offered a simple procedure to test for the addition of thickeners. If you thought you had some arrowroot-thickened "cream" on your hands, you needed "a few drops of a solution of jodine [iodine] in spirit of wine" to add to a sample of the cream. Real cream would turn yellow; fake cream would turn dark blue.[33]

Much of Accum's power to shock came from the way he subjected food to rigorous scientific analysis, and so he could not be dismissed as a scaremonger. At his Soho premises, he had a little sideline as a practising consultant on the adulteration of food: victimized members of the public could bring him samples of suspect food, and, like a Sherlock Holmes of the mustard pot, he would whip out his pipettes to give them dazzling proof of the crime. Accum claims in one of his books that he had twenty-eight years of experience in analysing the respective strengths of British beer and porter (the same kind of specialist knowledge as Sherlock Holmes's ability to spot 140 different types of tobacco ash). The *Philosophical Magazine* of 1819 told the story of how Accum had helped a poor charwoman solve the mystery of some strange blue tea.[34] This charwoman was in the habit of drinking green tea mixed with a teaspoonful of spirit of hartshorn, or ammonia, which she presumably added for medical reasons, since it can have done nothing for the taste (ammonia was sometimes taken internally to help circulation or relieve headaches). One day, this woman purchased an ounce of her customary green tea from a grocer's shop and made up a pot of tea, adding the hartshorn as usual. She was amazed at the "lively blue colour which the beverage made of it assumed." She took the tea back to the grocer, who seems to have professed ignorance. Puzzled, she took a sample of the tea leaves to Accum, who wasted no time in pronouncing the leaves to be coloured green with toxic copper, since copper mixed with ammonia goes bright blue. This he proved by mixing two parts of the leaves with one of nitrate of potash, throwing the mixture into a red-hot crucible. All that was left behind was copper, "in combination with the alkali of the saltpetre." Clearly, the so-called tea was actually some other kind of leaf—probably sloe leaves—tinted green with copper to look like real China tea. Thus were the lies of the grocer exposed and the humble charwoman vindicated.

The *Treatise* is filled with simple chemical tests that could be done to ascertain whether the food you had trustingly bought and brought home was real or fake. For example, Accum noticed that olive oil was often diluted with cheaper poppy seed oil. By freezing a sample of the oil, he could tell whether it was pure.[35] The olive oil would freeze, while the poppy seed oil would remain fluid. Equally, to tell whether lemonade had been doctored with tartaric acid, he recommended adding a concentrated solution of muriate of potash.[36] If a precipitate ensues, "the fraud is obvious." Few readers of Accum probably went to the trouble of obtaining and carrying muriate of potash samples around with them on the off chance that they should be offered a refreshing glass of lemonade. But the mere existence of these tests gave far greater potential power to the consumer than he or she had ever had before. Too often, a consumer in a pub complaining that a pint of beer was not pure would be met with brazen denial by the publican. Accum's chemical tests proved beyond doubt that the adulteration of beer was "not imaginary" but real, ranging from the relatively harmless additions of molasses and honey for sweetness and orange peel for fragrance to the more sinister admixtures of quassia and wormwood for bitterness, capsicum for pungency, and green vitriol, a substance to enhance the head of a pint of beer, giving it the sought-after "cauliflower" appearance. As a Westphalian, Accum took his beer very seriously and wrote another treatise, still popular among real-ale enthusiasts, on the most wholesome ways of making this essential beverage.

But to understand what made Accum's original *Treatise* so electrifying, we have to see it in the context of its times. Its appeal was partly due to the fact that it appeared at a point and place in history when, for the first time, adulteration on an industrial scale had become a serious and endemic problem. Accum himself quotes an anonymous source from 1773 saying that "Our forefathers never refined so much; they never preyed so much on each other; nor, I presume, made so many laws for their restraint, as we do."[37] Many of the deceptions Accum was describing were relatively new. It was only in an industrialized and impersonal city that the swindlers could get away with the crimes Accum described. Britain in 1820 had the most highly industrialized cities in the world, coupled with a relatively laissez-faire government that failed to police bad food in the way that it was policed in other industrial cities, such as Paris. As a result, adulteration affected the lives of everyone. One of Accum's reviewers described as "almost ludicrous" the extent to which the deceptions were carried on:

So inextricably are we all immersed in this mighty labyrinth of fraud that even the vendors of poison themselves are forced, by a sort of retributive justice, to swallow it in their turn. Thus the apothecary, who sells the poisonous ingredients to the brewer, chuckles over his

19

roguery and swallows his own drugs in his daily copious exhibitions of brown stout. The brewer, in his turn, is poisoned by the baker, the wine-merchant and the grocer.[38]

And yet they all continued in their business, as if nothing were the matter. This is why the story of adulteration starts with Accum, who, while half in love with the magnificent industrial activity of his chosen country, retained a Germanic dismay at the British failure to care enough about the way that food was falsified "to a most alarming extent in every part of the United Kingdom."[39]

Industrial Britain and the "Insatiable Thirst for Gain"

Accum depicts Britain in 1820 as an exciting but terrifying place, where anything could be bought for a price—even the tremulous foetuses of young cows, purchased by pastry cooks "for the purpose of making mock turtle soup"[40]—but where staples are pushed to the lowest prices possible, making padding and adulteration almost inevitable. This is a society intensely and foolishly class-conscious, with everyone aspiring to eat the white bread of the rich and to feed their children an array of multicoloured candies that would once have been the preserve of the wealthy, but where almost no one asks *how* their bread can be so cheap and yet so white, or *why* their children's sweets can be coloured in shades not known in nature. It is a country where cunning and ignorance combine to create hazardous patterns of eating. Accum conveys how easy it was for the unprincipled to debase food often already debased—when Lancashire dairies heated up milk in lead pans, and innkeepers in the north of England witlessly ground mint for "mint salad" using a giant ball of lead instead of a pestle and mortar so that "portions of the lead are ground off at every revolution of the ponderous instrument."[41]

To a certain extent, such ignorance was not new. Lead had been used in cooking since ancient times, as we shall see in the next chapter. The difference was that no one in ancient times had known that lead was poisonous, whereas by 1820 scientists had known of lead's toxic properties for over a hundred years. What shocked Accum was that as the elite of British society progressed in science and industry, general ignorance in the kitchen seemed to get worse, not better. The problem was partly that the British had already lost much of their heritage of peasant cooking, as land enclosures drove smallholders off the land. These enclosures had been taking place since the sixteenth century, but they escalated in Accum's lifetime; between 1750 and 1850, there were more than four thousand Acts of Enclosure. These deprived tens of thousands of country-dwellers of the ability to forage for wild greens and berries—as woodland got swallowed up into vast country estates—or to grow their own vegetables and keep their own chickens, as they had once done.[42]

After the end of the Napoleonic Wars, agriculture was in an even worse state, with a slump in prices leaving many farmhands either out of work or living on a miserably low wage. The skill of soup making, always a basic part of peasant knowledge, now declined. The cookery writer Eliza Acton, in 1855 looking back over the previous half-century, observed in 1815 that the English had lost "the art of preparing good, wholesome, palatable soups, without great expense."[43] Already, in *The Cook's Oracle* published in 1817, Kitchiner was complaining that when the English did make soup, they smothered it with spice. What had been forgotten was the art of the simple pottage made from roots. A cook who could not so much as make a basic soup was not likely to have the necessary knowledge to guard against the wily tricks of swindlers.

It was not that Accum found all English food bad. He thoroughly admired the English habit of eating a lot of fresh meat, and he credited this diet with "the striking fact that the English soldiers and sailors surpass all those of other nations in bravery and hardihood," an ingratiating comment that contrasts with his insistence on accuracy elsewhere.[44] From his recipe book entitled *Culinary Chemistry*, it is obvious that Accum enjoyed foraging for seasonal foods, picking nasturtium pods in July, red cabbage in August, and mushrooms in September. He adored the whole calendar of English "domestic fruits." He gives recipes for conserved gooseberries, greengages, damsons, peaches, nectarines, bullaces; for apricot paste and other lovely things.[45] As well as having a passion for making jams and jellies

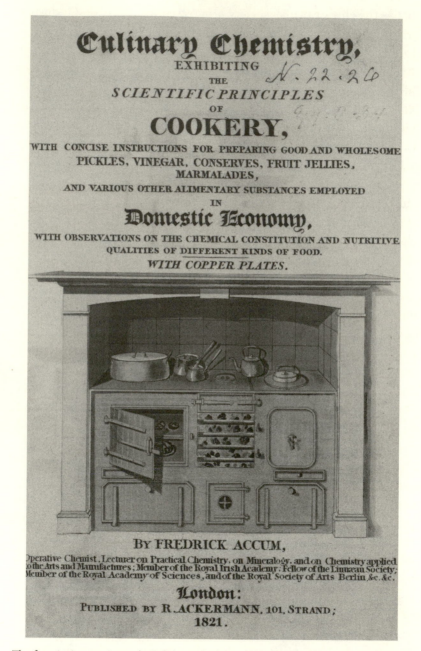

The frontispiece to Accum's *Culinary Chemistry* (1821), a book that showed him to be a connoisseur of good, simple food.

(always with punctiliously well sterilized jars), he wished to over-come the "vulgar prejudice" against turning these fruits into wine. English blackcurrants, he thought, were ideally suited to making a wine like "the best of the sweet Cape wines." Sloe and damson juice mixed with elderberry juice would make an approximation of port. Meanwhile, "grapes of British growth, are capable of making excel-lent sparkling and other wines, by the addition of sugar. I have made wine from immature grapes and sugar, which so closely resemble the wines called Grave and Moselle that the best judges could not distin-guish them from foreign wines."[46] High praise, from a German.

The main problem with British food, in Accum's opinion, was not the raw materials themselves but the dreadful things British cooks did with them. Accum reproved the English tendency to hurry over dinner yet waste "whole hours over the bottle as if time were of no value." He contrasts this with France, where a "good table" is "a grand object in life."[47] For a coffee-addict such as Accum, it was evidently distressing to live among people who had never learned the basics of coffee making. Coffee, Accum believed, "diffuses over the whole frame a glow of health, and a sense of ease, and well-being which is extremely delightful."[48] The English version, however, simply made him miserable. Most of what passed for coffee in England was little more than bitter "coloured water," he grumbled—hardly surprising given the ubiquity at grocer's shops of "sham coffee," made from burnt peas and beans.[49]

Even when real coffee was used, it often tasted terrible. The standard advice was to boil the coffee for five minutes, then boil for another another five minutes with isinglass (a clearing agent made from the bladder of sturgeon), before leaving it to stew for a further ten min-utes until it was soot-black and acrid.[50] Accum much preferred fresh strong coffee roasted to a "deep cinnamon" colour and brewed for a minimum amount of time. He bought his own beans, roasted them, and ground them himself in his own coffee mill, following the advice of a retired grocer that you should "never, my dear fellow, purchase from a grocer any thing which passes through his mill."[51] As for brew-ing, he used the modern percolation method invented by the Ameri-can scientist Benjamin Thompson, Count Rumford (1753–1814).

A coffee pot invented by Accum's hero, Benjamin Count Rumford.

Accum agreed with Rumford that "Coffee may easily be too bitter, but it is impossible that it should ever be too fragrant"; not a lesson that the English had learned.[52] Accum also agreed with Dr. Johnson that "he who does not mind his belly will hardly mind anything else."[53] He writes, a little wearily, of the way that melted butter forms the basis of almost every English sauce—"melted butter & oysters, melted butter & parsley, melted butter & anchovies, melted butter & eggs, melted butter & shrimps, melted butter & lobsters, melted butter & capers." You can tell he has himself endured one too many plates of meat swimming in butter at grand English tables on his way up the social ladder. He is impatient of English snobs, citing an absurd "Lord Blainey" who claims that hams "are not fit to be eaten unless boiled in Champagne."[54]

While the rich boiled their hams in champagne, the basics of a good diet were neglected. People judged food too much on what it looked like, and not enough on what it tasted like, or whether it did their bodies any good. Bread was the capital example. In "this metropolis," wrote Accum, bread was "estimated entirely by its whiteness."[55] According to the "caprice of the consumers," white bread was good bread. Yet it was almost impossible to produce properly white bread, "unless the very best flour is employed," which was too expensive for most pockets. Therefore, to please their customers, bakers would take low-grade flour and "improve" it by adding the bleaching chemical alum, which could make the bread whiter, lighter, and more porous. Without the alum, the bread would remain moister longer, but have a "slight yellowish grey hue," like greying laundry, which put people off. Accum believed that the use of alum was almost universal in London at this time. Nor did he blame the bakers as much as he blamed other swindlers (partly because he thought that alum was relatively inoffensive compared to copper or lead). "I have been assured by several bakers, on whose testimony I can rely, that the small profit attached to the bakers' trade, and the bad quality of the flour, induces the generality of London bakers to use alum in the baking of their bread."[56] They got away with it because commercially minded Londoners did not expect them to do otherwise.

Accum was more exercised by the lackadaisical attitude of the British towards the safety of what they ate. The use of copper to heighten the green look of vegetables was a case in point. "Will it be believed," he exclaimed, "that in the cookery books which form the prevailing oracles of the kitchen in this part of the island, there are express injunctions to 'boil greens with halfpence or verdigrise,' in order to improve their colour!"[57] In his *Treatise on Adulterations*, Accum reproduces a sinister recipe from a cookbook called *Modern Cookery* for something called "greening," a mixture of verdigris, vinegar, alum, and bay salt to be used with "whatever you wish to green." In another book called *The Ladies Library*, Accum had found a recipe for pickled gherkins that recommended boiling the vinegar in a copper pot.[58] In a third book, Mrs. Raffald's well-known *The English Housekeeper*, there were directions to boil pickles with halfpence "or allow them to stand for twenty-four hours in copper or brass pans."

The use of copper to make pickles green was a case where commercial manufacturers were following the bad practice of the home cook. Accum regretfully noted that in order to sell well, pickled vegetables—whether gherkins, samphire, French beans, or green capsicums—needed a "lively green colour." The consequences of this taste for green pickles could be "fatal," as Accum lamented. He quotes a Dr. Percival on the case of "a young lady who amused herself, while her hair was being done, with eating samphire pickles impregnated with copper. She soon complained of pain in the stomach; and in five days, vomiting commenced, which was incessant for two days. After this, her stomach became prodigiously distended; and in nine days after eating the pickles, death relieved her from her suffering."[59] Copper was used in sweet things, too, "such as small green limes, citrons, hoptops, plums, angelica roots," always conveying an impression of vitality that was entirely misleading.[60] It was Accum's opinion that

> Copper cooking utensils are attended with so much danger, that the use of them ought to be laid entirely aside. They have not only occasioned many fatal accidents (which have been made public) but they have injured the health of great numbers, where the slower, but not less dangerous effect has not been observed. If not kept very clean and bright, the rubbing or scraping that takes place when making stews, or cooking dishes that require stirring and remaining a considerable time on the fire, always wears off some of the metal which impregnates the food, and has a deleterious effect.[61]

Copper-green foods were—alas—not the only example of reckless stupidity in the kitchen. This stupidity, though it was not ill-intentioned, could be every bit as damaging as the venality of real swindlers. "For many years," Accum wrote, British cooks had been using the leaves of the cherry laurel (*prunus lauro-cerasus*) to flavour custard. The reasoning was that cherry laurel leaves, when steeped in milk, give it a nutty flavour not unlike the much more expensive bitter almond. As a result, cherry laurel was used to communicate "a kernel-like flavour" to custards, puddings, creams, blancmanges, and

other delicacies of the table. There was just one flaw with cherry laurel. It was poisonous, and it had been known to be poisonous since
1728, when "the sudden death of two women, in Dublin, after drinking some of the common distilled cherry laurel water, demonstrated
its deleterious nature."[62]

Yet still cooks used it, believing it to be harmless in small quantities. As Accum remarked in a different context, "different constitu-
tions are differently affected by minute quantities of substances that
act powerfully on the system."[63] He cites a case from just the previous
year, 1819, when some children at a boarding school in Richmond
became "severely ill" after eating some custard flavoured with cherry
laurel. A girl aged six and a boy aged five fell into a profound stupor.
Two other girls developed intense gastric pain. It took all the children three days to recover. Such pain just for eating a custard! It is
impossible not to share Accum's rage at this dangerous folly. "What
person of sense or prudence, then, would trust to the discretion of an
ignorant cook, in mixing so dangerous an ingredient in his puddings
and creams? Who but a maniac would choose to season his victuals
with poison?"[64]

Accum gives the answer himself. Who but a maniac? A villain.
There were plenty of them about, judging from the evidence of the
Treatise, "unprincipled modern manufacturers" who hid under the
guise of respectability. Accum was angry enough with the "inexcus-
able negligence" of those cooks who poisoned eaters by mistake by
underestimating the dangers of copper pans or cherry laurel, but to
poison cunningly and knowingly was a crime akin to murder, driven
by naked greed. "The eager and insatiable thirst for gain, which seems
to be a leading characteristic of the times, calls into action every human
faculty, and gives an irresistible impulse to the power of invention;
and where lucre becomes the reigning principle, the possible sacrifice
of even a fellow creature's life is a secondary consideration."[65]

Such men would risk even the health of children, if it meant
enhanced profits. One of the most chilling chapters in Accum's *Treatise* is the one concerning "Poisonous Confectionary," in which he
writes of the "grossest abuses" in "those sweetmeats of inferior quality frequently exposed to sale in the open streets, for the allurement

of children."[66] Accum writes about these sweets in the same fearful terms that modern mothers might use to warn children against strangers. Children were tempted to buy these candies by the bright colours, colours that would do their little tummies no good.

> The white comfits, called sugar pease, are chiefly composed of a mixture of sugar, starch, and Cornish clay (a species of very white pipeclay;) and the red sugar drops are usually coloured with the inferior kind of vermilion. The pigment is generally adulterated with red lead. Other kinds of sweetmeats are sometimes rendered poisonous by being coloured with preparations of copper.[67]

After Accum's *Treatise* appeared, a father wrote a letter to *The Times* saying that his family had suffered terrible stomach pains, sickness, and retching after eating some multicoloured comfits the shape of "turnips, parsnips, carrots and beans" bought as a treat.[68] Some treat!

Yet the confectioners who sold such poison candy often pleaded ignorance of their side effects; others admitted that lead or copper was bad but claimed that they did not know that these poisonous additives were present in their particular product; and they weren't *always* lying. In the world described by Accum, no single person can take responsibility for the quality of a given food or drink, since it has passed through so many hands. Adulteration thrives when trade operates in large, impersonal chains. In a rural setting, swindling is a risky business. If you are the village milkman, the chain between you and your customers is very short: you know them all by name because they are your neighbours. If you start watering down your milk, the chances are that word will soon get out and you will be ostracized. But if you are selling milk in the metropolis of London in 1820, to an ever-shifting clientele, it is easier to cover your tracks. With less perishable goods, such as tea, sweets, or spices, it is easier still. The chain between producer and consumer may be so long that it becomes hard to determine who exactly in the chain did the dirty deed.

Sometimes, indeed, the chain was so long that food could become poisoned without anyone intending it. Accum noted that during "the long period devoted to the practice of my profession," he had had "abundant reason to be convinced that a vast number of dealers, of the highest respectability, have vended to their customers articles absolutely poisonous, which they themselves considered as harmless, and which they would not have offered for sale, had they been apprised of the spurious and pernicious nature of the compounds."[69]

The most startling case Accum cited was of some Double Gloucester cheese, which through a convoluted sequence of events became coloured with red lead. Ordinarily, Double Gloucester—so called because it originally used the milk from two milkings—is coloured, when it is coloured at all (sometimes it isn't), with annatto. Annatto (now known as E160 (b)) is a vegetable dye that comes from an orange pulp on the tropical annatto tree. A few people are allergic to it, but otherwise it is fairly harmless. Red lead, on the other hand, is lethal. Accum repeats a story told by a Mr. J. W. Wright of Cambridge of a gentleman who had reason to stay for some time at an inn in a city in the West Country. One night, he was "seized with a distressing but indescribable pain in the region of the abdomen and the stomach, accompanied with a feeling of tension, which occasioned much restlessness, anxiety and repugnance to food." Twenty-four hours later, he was completely better. Four days later, exactly the same thing happened—the agony, the tension, the anxiety, the recovery. Now the gentleman remembered that on both occasions he had asked the mistress of the inn for a plate of toasted Gloucester cheese, a dish he often ate for supper at home. The mistress was affronted at this news. Why, she had purchased the cheese from a respectable London dealer. But when the gentleman ordered the toasted cheese for a third time, and for a third time suffered "violent cholic," there was no mistaking it; the cheese was to blame. Then a serving maid chipped in that a kitten had been violently sick after chewing a rind of the same cheese.[70]

At this, the mistress swallowed her pride and had the cheese examined by a chemist, who pronounced it contaminated with red lead. The respectable London dealer now asked the farmer who had manufactured the cheese how it had become contaminated; he said he had bought the annatto from a mercantile traveller who had supplied him for years without causing "a single complaint"; but he in turn had got it from another supplier who had touched up the

29

annatto with vermilion (a nonpoisonous dye), which had been mixed with red lead by a druggist who had assumed the vermilion was only to be used as "pigment for house painting," never imagining it would go into cheese. Thus, concludes Accum's source, "through the circuitous and diversified operations of commerce, a portion of deadly poison may find admission into the necessaries of life, in a way which can attach no criminality to the parties through whose hands it has successively passed."[71]

Often, though, adulteration arose by exploiting the division of labour in a much more conscious way. Given his admiration for industry, it upset Accum to see how the adulterators mimicked "the order and method of a regular trade," in which sinister kingpins would manipulate workmen who were often "ignorant of the substances which pass through their hands."[72] "It is a painful reflection, that the division of labour which has been so instrumental in bringing the manufactures of this country to their present flourishing state, should have also tended to conceal and facilitate the fraudulent practices in question; and that from a correspondent ramification of commerce into a multitude of distinct branches, particularly in the metropolis and the large towns of the empire, the traffic in adulterated commodities should find its way through so many circuitous channels, as to defy the most scrutinizing endeavour to trace it to its source."[73] The swindlers did everything possible to evade capture:

> To elude the vigilance of the inquisitive, to defeat the scrutiny of the revenue officer and to ensure the secrecy of these mysteries, the processes are very ingeniously divided and subdivided among individual operators, and the manufacture is purposely carried on in separate establishments. The task of proportioning the ingredients for use is assigned to one individual, while the composition and preparation of them may be said to form a distinct part of the business, and is entrusted to another workman. Most of the articles are transmitted to the consumer in a disguised state, or in such a form that their real nature cannot possibly be detected by the unwary.[74]

Take the liquor business. Dodgy liquor manufacturers were in the habit of using something called extract of *Cocculus indicus* (made by boiling the berries of that plant) to make porter and ales extra intoxicating. But to cover their tracks, the market name for this substance was "black extract," the implication being that it would be used by tanners and dyers, rather than by liquor-makers.

Some forms of adulteration were crude—such as "P.D." or pepper dust, a "vile refuse" swept from the floor of the pepper warehouses that often got mixed in with ground black pepper. An even worse version was known as "D.P.D.," short for "dust of pepper dust," the very grimiest, nastiest floor sweepings of all. There was little art in this deception. On the other hand, some truly intricate labour went into many of the more outrageous forms of adulteration. Real black peppercorns—which many consumers doubtless bought in preference to ground, thinking, wrongly, that there was no way to adulterate the whole spice—were sometimes padded out with "factitious peppercorns," which seem to have been the work of an artisan to make. First, blackish "oil cakes" were taken (the residue left over after pressing linseed oil) and mixed together with common clay and some cayenne pepper (to give the "corns" some bite so that consumers might really be fooled by them). Then this paste was pressed through a sieve and rolled in a cask until it formed little pellets. Making these tiny balls of fake "pepper" must have been highly laborious, and swindlers couldn't get away with adding very many to the real peppercorns before suspicion was aroused—around 16 percent was standard—yet evidently it was still worth the swindlers' while to do it. Labour was cheap and spice was expensive (the duty on a single pound of pepper was 2s. 6d., a tax that was reduced in 1823), and there was no shortage of workers to carry out this peculiar trade.[75]

It was in the manufacture of fake tea that the intricacy of the adulterator's art reached its pinnacle. Accum details numerous cases where "tea" was not tea at all, but elder leaves, ash leaves, or, most often, sloe leaves, boiled and baked and curled and dried and coloured until they resembled the best China green tea. In 1818, the attorney general brought numerous cases against tea swindlers, including one against a grocer, Mr. Palmer, who had been purchasing imitation tea from a pair of crooks called Proctor and Malins. These crooks engaged another man, Thomas Jones, to gather the sloe leaves and whitethorn leaves for them. Jones seems to have been unaware, at least at first, what the leaves were for. He in turn subcontracted the spadework of picking the sloe leaves to another man, selling the leaves on to Proctor and Malins for tuppence a pound.[76]

Now the real villainy began. To convert the leaves into "an article resembling black tea," they were first boiled, then sifted to separate out the thorns and stalks, and baked on an iron plate to dry out, then "rubbed with the hand, in order to produce that curl, which the genuine tea had." It was then coloured with logwood (*Haematoxylon campechianum*), a dye from the West Indies, which in large doses

Accum's illustration of real and fake tea leaves, from his *Treatise on Adulterations* (1820). Those on the left with "deeply serrated" edges are real. Those on the right are sloe leaves.

can cause gastroenteritis. The "green tea" was more poisonous still, being boiled with verdigris (copper acetate, a poison) and, when dry, painted with a toxic mixture of "Dutch pink" dye and more verdigris. To bring the scandal of this to life, the crown prosecutor told the jury that whenever they supposed "they were drinking a pleasant and nutritious beverage, they were, in fact, in all probability, drinking the produce of the hedges round the metropolis, prepared for the purposes of deception in the most noxious manner." Needless to say, the jury found the defendants guilty.

This case—along with the other tea prosecutions of 1818—demonstrated how widespread tea adulteration really was. In his book, Accum included sketches of sloe leaves and tea leaves to show how differently shaped they were. Tea leaves, he noted, were "slender and narrow," while sloe leaves were round and "obtusely pointed," with less serrated edges.[77] Once rolled up and painted with verdigris, though, the differences would be much harder to spot—unless you brought the tea home and moistened it and rubbed it against a white sheet of paper, as Accum suggested, to bring off the dye; but by this point you would already have been swindled. Even a high price was no guarantee that tea was genuine. Accum himself managed to obtain twenty-seven different samples of "spurious tea" from the merchants of London, ranging from "the most costly to the most common." Tea fraud was thus a problem affecting the whole of British society. For much of the previous century, tea had been seen as a rather feminine and aristocratic drink; but now the habit of tea drinking had spread to the working-class masses. Heavily sweetened tea sustained the workforce that powered Britannia's industrial advance. To poison that workforce with sloe leaves and verdigris was a serious matter. What should be done about it, though, was a matter of fierce disagreement, because it touched on the most divisive political issue of the day, the question of the relationship between government and commerce, law and freedom. How far could legislation interfere with the divine liberty of free trade?

Adulteration and the Law

Reading British newspapers from this period, you often come across the view that, while adulteration was regrettable, it was a natural consequence of the free trade that was necessary to power Britain's commercial success. To introduce new regulatory measures would have the undesirable effect of stifling the market. It was therefore better to do nothing. *Laissez faire, laissez aller, laissez passer*, as the Physiocrats said: keep government out of trade. Accum observed that "public opinion" now saw adulteration as just another "mercantile pursuit," and it was thus "not only regarded with less disgust than formerly, but is almost generally esteemed as a justifiable way to wealth."[78] He summarized the commercial defence of adulteration:

> It has been urged by some that, under so vast a system of finance as that of Great Britain, it is expedient that the revenue should be collected in large amounts; and therefore that the severity of the law should be relaxed in favour of all mercantile concerns in proportion to their extent: encouragement must be given to large capitalists; and where an extensive brewery or distillery yields an important contribution to the revenue, no strict scrutiny need be adopted in regard to the quality of the article from which such contribution is raised, provided the excise do not suffer by the fraud.[79]

The rule of *caveat emptor*—let the buyer beware—also came into play. If consumers chose to buy falsified foods, that was their lookout—an argument we will return to in chapter 3. To Accum, familiar with the much more interventionist system of Westphalia, such arguments were unconstitutional and unjust. The "true interests of the country" dictated that adulterators, both large and small, be dealt with by the strong arm of the law. As for the supposed loss in tax revenue that would come from exposing them, Accum believed that "a tax dependent upon deception must be at best precarious." His strongest reasons for opposing this argument, though, were moral. To paint sloe leaves with poison and pass them off as tea was wicked; and far from turning a blind eye, the law should crack down on such activity. His indignation roused Accum to one of his most dramatic statements: "It is really astonishing that the penal law is not more effectually enforced against practices so inimical to the public welfare. The man who robs a fellow subject of a few shillings on the high-way, is sentenced to death; while he who distributes a slow poison to a whole community, escapes unpunished."[80] This wasn't quite right; the last British citizen to hang for highway robbery had been executed in 1783. Accum had made his point, though. Too many swindlers were getting away with it, and the law needed to intervene. Where government revenue was at issue, con men could be prosecuted for adulterating taxable items such as beer or tea. But there was no specific law against adulteration, and little consistency in the way that the existing law was applied. More government action was needed to extend penalties "to abuses of which it does not now take cognisance"; that enlarging the law in this way, Accum argued, would actually benefit 'the revenue."[81]

Not everyone shared his opinion, though. Some thought "the revenue" was to blame, in the first place, for much adulteration. If imported goods weren't taxed so highly by a greedy government, there would be no motive for making cheap imitations from native plants. *The Times*, whose general outlook was favourable to laissez-faire, responded to the fake tea scandals of 1818 in exactly the opposite way to Accum. What was needed was not more government interference but less. It shared Accum's disgust at the tea counterfeiters, remarking that many took refuge in the "mild refreshment" of tea because they feared the "deleterious drugs" used in porter. What an injustice it was that "sober men and women are to be poisoned as a penalty for refusing to get drunk." But unlike Accum, *The Times* was not altogether surprised that nine-tenths of the "tea dealers of the metropolis" sold the "adulterated article in a greater or less proportion." An anonymous editorial written in March 1818 held that adulterated tea was

inevitable—inevitable, so long as the genuine tea of China shall be taxed very nearly one hundred per cent on its original cost and sold to the people of England, who are the greatest consumers of it in the world, at a higher price than to any of those nations who never send a single merchant ship to sea. We are not so childish to adjure a financier by his regard for the health or enjoyments of his countrymen, to reduce a tax on a simple article of consumption; but we would appeal to the common sense and plain arithmetic of our law-givers, whether it be for their interest to maintain the tea-duties at such a rate as to supersede the demand for this popular article, and to extinguish the revenue desirable from it, by forcing into the market a spurious commodity, which is only advantageous to manufacture, because of the exorbitant imposts on tea. Mr Pitt knew better.[82]

This last was a reference to the fact that when Pitt the Younger was prime minister (1783–1801), he had slashed tea duties from 119 percent to 25 percent in order to combat smuggling. It had been a very successful policy, vastly increasing the amount of tea passing through the Exchequer. These were different times, though, with the price of all food at a premium following the wars with France. The Corn Laws, introduced in 1815 to safeguard the livelihood of British farmers, kept the price of wheat, and thence bread, artificially high. Meanwhile, duties on luxuries such as tea, wine, spirits, and tobacco were all extremely high in 1820, which doubtless contributed to the market in counterfeit versions.

Accum himself makes frequent reference to the effects of "the late French war" on food and drink, especially beer. During the Napoleonic Wars, many lines of trade were shut off, pushing up the price of all raw materials. London porter, so Accum noted, was "unquestionably" weaker at this time, except when it had been made "intoxicating" by the addition of "narcotic substances."[83] He describes the special jackets worn by brewers with many large pockets, each concealing a different adulterating substance—spices, colours, and drugs.[84] During the wars, Accum observed, imports into Britain of a drug called *Cocculus indicus* shot up. This is a convulsive poison, whose active ingredient is picrotoxin, a powerful and bitter narcotic, now used to destroy lice or stupefy fish. In those days, it was added by fraudulent brewers to beer, to make its effects stronger and cover up the fact that not enough malt and hops had been used. Brande's *Manual of Chemistry*, published in 1819, noted its dangers—a fact seemingly known to the brewers themselves, though it didn't stop them from using it. In *Every Man His Own Brewer* (1790), Samuel Child admitted that *Cocculus indicus* was "poisonous leading to stupefaction, and unlawful; being of excessive strength to attack the head"; yet he still includes it in his recipe for porter.[85]

The adulteration of beer was not only an attack on drunkards. Since so much water, especially in cities, was undrinkable, beer had

"Death's Register" by Richard Dagley from *Death's Doings* (1826), showing a skeleton plotting various poisonous swindles. Notice the sign on the wall, which reads "Accum's List" and the barrel of *Cocculus indicus*, which was an adulterant of beer.

the status of a basic family drink, consumed by children as well as adult men and women. This made tampering with it a grave matter, affecting the whole population. The problem was widespread. Accum cites countless prosecutions against brewers and publicans for adulterating beer in one way or another, or for mixing strong beer with weaker "table beer." Between 1813 and 1819, more than thirty brewers were fined vast sums for "receiving and using illegal ingredients in their brewings." Among them, John Cowell was fined £50 for using Spanish liquorice and mixing table beer with strong beer; John Gray was fined £300 for using ginger, hartshorn shavings, and molasses; Allatson and Abraham were fined £630 for using *Cocculus indicus, multum,* and "porter flavour."[86]

What these prosecutions reveal, though, is that, at least as regards beer, the government intervened more to protect the consumer than Accum allowed. British parliamentary laws safeguarding beer at this time were comparable to the notorious *Reinheitsgebot* of Bavaria, promulgated in 1516. Acts of Parliament prohibited beer from containing any substance but malt, hops, and water (yeast was added to beer only later).[87] Druggists and grocers could be prosecuted for selling adulterating items to brewers, even something as harmless as molasses (a dark treacle used to colour and sweeten beer). Even the use of burnt sugar to add colour and isinglass to clarify beer were prohibited by law.[88] Never again would the standards be so stringent.

Indeed, the standards for British beer now are nowhere near as strict as they were then. Practically all beer sold in modern British pubs would count as "adulterated" in Accum's view, even if not adulterated with such noxious substances as *Cocculus indicus.* By law British beer is permitted to contain caramel to adjust its colour, potassium chloride to adjust its flavour, phosphoric acid to change its acidity, and numerous other additives and processing aids. Brewers may choose from no fewer than seventeen different preservatives, such as calcium sulphite or sodium benzoate.[89] To Accum, most of these substances would have been needless deceptions; and the law agreed with him. The case of beer shows how inconsistent the British state of 1820 was in its laissez-faire attitudes toward food and drink. While it chose not to interfere in cases of food adulteration, even poisoning, it could be very strict in regard to beer, a remnant of older patterns of government regulation of what people ate and drank.

Clearly Accum was writing at a crossroads, as the new, impersonal modern trade in food superseded older attitudes. Urbanization, as one historian has said, "had deprived millions of people of experiences that had been commonplace for humankind before"—the taste of honey, for example; not long after Accum, a London street sweeper remarked that he had never tasted honey but had heard it was "like butter and sugar mixed."[90] Accum rightly recognized a new world in which the integrity of most basic foods was constantly under attack; this is the world we still live in today. But, unlike most of us now, Accum remembered enough of the old world to remember what unadulterated food should really be like. He knew the "genuine old . . . beer of the honest brewer" with its "rich, generous, full-bodied taste."[91] And he knew it was a disgrace that instead of this nourishing drink, so many people were making do with a product that was stale, coloured, or drugged.

Accum's Disgrace

Given the uproar that surrounded Accum's *Treatise*, you might expect that it would have led to some change in British law. It did not do so. The "Death in the pot" campaigners would have to wait another forty years before the first anti-adulteration legislation was passed, in 1860. There were various reasons for this.

The prevailing mood of the British press and government in 1820 was still excessively laissez-faire: nonintervention remained the dominant philosophy of the day, and it took several decades for this to change. Only in the second half of the nineteenth century, after years of pressure from socialists, Chartists, and radicals, did the doctrine lose its dominance.[92] Accum was thus something of an anomaly in his day: a man of trade who favoured state intervention.

Another problem was the state of the science: Accum's chemical tests, though advanced by the standards of 1820, were still somewhat limited and would be superseded, first by the more comprehensive chemical tests of French chemistry, and later by the microscopic food analysis of the 1850s. But there was also a third reason, which had to do with Accum himself. If 1820 was the year of his greatest triumph—a year when he was feted by the press and had a glowing profile written of him in *European Magazine* that hailed him as the "most popular consulting chemist" in Britain—it was also a year in which he lost everything and was to flee the country with his reputation ruined. As a result, Accum never got the chance to press home the legal questions that his *Treatise* had raised. For Accum, the scourge of swindlers, was caught out in a swindle of his own.

On the 5 November 1820, Guy Fawkes Day, the assistant librarian at the Royal Institution, a Mr. Sturt, discovered that several books in the Reading Room had been mutilated—various leaves and plates had been torn out.[93] Since these books were ones that Accum was in the habit of reading, the finger of suspicion pointed immediately at him. His connection with the Royal Institution had been important to Accum's career. So highly did he regard the institution that he dedicated his first book on chemistry to its managers. After he gave up working there, he continued to be a subscribing member, which gave him rights to use the Reading Room, with its extensive library of books. For an intellectual magpie like Accum, it must have been an invaluable resource; though, judging by later events, it was one whose limits he failed to understand.

After Accum left the Reading Room on 5 November, the assistant librarian went over to look at the books he had seen in Accum's hand. Sure enough, some leaves had been torn out. To make sure that the crime really had taken place, the secretary of the institution ordered Mr. Sturt to have peepholes bored in a cupboard adjoining the Reading Room. From there, it was possible to spy on the activities of those using the library. On 20 December, Accum fell into the trap. From the safety of his cupboard, Sturt observed Accum tearing several pages out of a volume of *Nicholson's Journal*, before leaving the institution in a "hurried and confused manner." A search warrant was put out. After Accum's Old Compton Street residence was ransacked, the Royal Institution librarian identified a further thirty torn pages on his premises. Accum's defence was that he had torn them from his own books; but he was arrested and charged with robbery. This case was dismissed by a magistrate who was clearly on Accum's side, remarking that once separated from the books, the leaves just amounted to "waste paper," of no value. Needless to say, this did not satisfy the bigwigs of the Royal Institution. They brought a new prosecution against Accum of "feloniously stealing and taking away two hundred pieces of paper of the value of ten pence."[94]

Having been the toast of the town only a month earlier, Accum was now a social embarrassment and on his way to becoming a pariah. The newspapers made fun of him. A feeble verse called "Death in the Pot" circulated in the press:

> What is his crime? A trick at most,
> A thing not worth debating.
> 'Tis only what the *Morning Post*
> Would call *Accum*-ulating.[95]

After the scandal, he was dropped unceremoniously by his publishers, Longman, Hurst, Rees, Orme & Browne, despite his astonishingly prolific output and healthy sales. But other friends rallied round. He was accompanied in court by John Papworth, an architect, and Rudolph Ackermann, the publisher, who paid £100 each in "surety" to keep Accum out of jail.

His friend Anthony Carlisle wrote to *The Times* begging Earl Spencer, president of the Royal Institution, to reconsider and stop this "*persecution* against a man of science." Carlisle did not deny that Accum was guilty of "gross misconduct," but he argued that Accum had learned the habit of plundering books in this way from the scientist William Nicholson, who would often tear up books "to save time and trouble." Carlisle insisted that Accum and Nicholson looked on books as no more valuable than "crucibles" or even pots and pans. He pleaded, moreover, that because of his lack of a "refined education" (reading between the lines, this meant because he was un-English, and more particularly of Jewish blood), Accum did not see the full "moral turpitude" in his actions. Carlisle regretted that Accum appeared unable to make "a competent apology" for himself but begged His Lordship to reconsider, praising Accum as a "remarkably ingenious man," a scientist of "native simplicity, lively intelligence and frankness."[96]

It was hopeless. The prosecution went ahead on 5 April 1821. *The Times* haughtily commented how generous the Royal Institution had been in allowing Accum to remain "at large on bail."[97] But Accum

41

did not appear at the hearing, having left for his native Germany, unable to bear the shame and loss of popularity any longer. He would never return to England. Aged fifty-three, he was forced to rebuild his life, becoming a professor of technical chemistry and mineralogy at the Royal Industrial Institute in Berlin. He would die in Berlin in 1838 at age sixty-nine, having completed a final work in German not on adulteration but on the properties of building materials. He seems never to have got over the humiliation of being ostracized by English society. After 1820, his English works were published either anonymously or under the pseudonym "Mucca," a reversal of Accum, whose sound underlined the fact that his name was now muck. Seldom has anyone moved from ubiquity to obloquy so fast. In 1822, in the course of a satire on the press, the poet John Hamilton Reynolds (1794–1852) asked: "Mr Accum—ay, what has become of Mr Accum?"[98] He had become no more than a footnote.

England was the poorer for Accum's absence, not least because he never got the chance to continue his work on adulteration. One of the most controversial aspects of his *Treatise* had been the way he fearlessly named and shamed individuals for their part in food fraud, using minutes from parliamentary committees of investigation to expose to the glare of publicity "respectable" villains who might otherwise have covered up their activities. In his second edition, he promised to continue these exposés. The book-mutilating scandal put paid to that. Some have been tempted to see a conspiracy in Accum's disgrace. Was he framed by powerful interests who did not want him pressing home his investigations? The historian of science C. A. Browne wrote of Accum's "secret enemies," arguing that "the work of the reformer in all ages has been rewarded with hatred and abuse."[99] In the second edition to the *Treatise*, which appeared in April 1820, Accum himself hinted at secret battles. His practice of naming and shaming swindlers had given rise to threats against him, presumably from manufacturers. He wrote in reply:

> To those who have chosen anonymously to transmit to me their opinion concerning this book, together with their maledictions, I have little to say; but they may rest assured, that their menaces will in no

way prevent me from endeavouring to put the unwary on their guard against the frauds of dishonest men, wherever they may originate; and those assailants in ambush are hereby informed, that in every succeeding edition of the work, I shall continue to hand down to posterity the infamy which justly attaches to the knaves and dishonest dealers who have been convicted at the bar of the Public Justice of rendering human food deleterious to health.[100]

In other words, had Accum not been ruined, he would have devoted his energies to continuing the fight against adulteration. Was he therefore prevented from doing so by malign forces?

Probably not. Even though he had powerful enemies, they were merchants and not scientists; the humble assistant librarian of the Royal Institution seems to have had no motive for lying about what he saw through the peephole. Anyway, the book-mutilating incident seems entirely in keeping with what we know of Accum's character. Even his friends admitted that he was both impetuous and impatient and saw books simply as a means to the end of advancing his own cause. It takes little familiarity with Accum's work to see that he engaged in repeated low-level literary plagiarism. He had a habit of lifting whole passages unacknowledged from other writers. In one instance he even lifted a whole section from one of the reviews of his *Treatise*, to use as his own in writing another book. Sometimes, he would quote a source, then continue the quotation without quotation marks, leading the reader to suppose they were reading Accum's own prose. He did this with Count Rumford's thoughts on coffee in his *Culinary Chemistry*. This seems perhaps less dishonest than clumsy, since Accum did quote from Rumford on the same page. Much of Accum's writing was effectively anthologizing, and perhaps he sometimes forgot where the anthology stopped and his own thoughts began.

The example of Accum illustrates the extent to which literary fraud is endemic, both in science and in cookery. Hannah Glasse, the best-known English cookery writer of the eighteenth century, stole 263 of her recipes from another source, yet her reputation remains high among culinary historians. Accum was not so lucky. The insertion of someone else's ideas into your own is a form of adulteration.

43

As much as he hated the adulteration of food, Accum was himself a literary adulterator. Perhaps this was what enabled him to understand the mindset so well.

Then again, it is not clear how much of this literary theft was premeditated, and how much the result of taking on too much work. In 1820, Accum was pushing himself ridiculously hard. That year alone, he had brought out no fewer than three books on three different subjects, not counting revised editions; the previous year, he had published two more; and a further two were in the pipeline. There must have been moments when note-taking by hand seemed too laborious for the task. Easier to tear the pages out for instant reference. Can we blame him? Had he lived in the era of photocopying, or employed a team of researchers, as his modern equivalents do, Accum would surely never have felt the need to vandalize his source books.

Even if no conspiracy, Accum's disgrace undoubtedly set back the battle against the swindlers before it was properly started. Without his energy, the cause stalled. And even in the details of his disgrace, there is proof that he intended to continue his fight against adulteration. The work that Accum was seen stealing from the volume of *Nicholson's Journal* on that fateful day in December 1820 was Parmentier's essay "On the Composition and Use of Chocolate." Chocolate was not included in Accum's original treatise; he was evidently considering its inclusion in future editions. Parmentier, a French scientist, was a man after Accum's own heart, who saw chocolate as both a medicine and a food. In this essay, Parmentier praised chocolate as "agreeable" but warned against the "frauds" committed in its fabrication.[101]

Chocolate—which was taken then as a drink rather than in bar form—was often padded with floury substances or spoiled in some other ways. Parmentier warned against chocolate that left a "pasty taste in the mouth"; chocolate smelling of glue; and chocolate that jellified as it cooled down, for all these were signs that "farinaceous matter" had been added. If it smelt of cheese, animal fats had been added. If it left grainy deposits behind, the cocoa beans had been either badly picked or mixed with poor-quality raw sugar. If it tasted bitter, this was a sign that the cocoa used was too green; if musty, the cocoa was "decayed."[102] We can picture just how interesting this would have been for Accum; so interesting that he tore it out instead of transcribing it. Had he not been arrested, might he have set to work analysing the chocolate of London, sniffing it for cheesiness, tasting it for bitterness, scrutinizing it for floury deposits? We will never know.

Accum was a huge loss to the fight against swindling. He was not just the first systematic campaigner against adulteration, he brought a range of personal qualities to the cause that have never been matched. He may have been a flawed scientist and a plagiarist, arrogant and messy in his methods, but his many achievements dwarf the scandal that ended his career. As well as chemical brilliance, he had enough charisma to carry the public along with him (unlike later scientists, who have sometimes had difficulty persuading consumers that cleaning up the food supply is important). As well as a sense of fun, he had a great moral seriousness—a proper disgust at "death in the pot" that made him politically fearless. By focusing on swindles that were poisonous rather than those that were merely cheats, he gave the cause urgency. Above all, Accum had a great passion and feeling for food. Most of the "pure food" evangelists who came after him got mixed up in sterile notions of purity. Accum never made this mistake. He never threw out pleasure along with poison. He never forgot that adulteration was an attack, not just on people, but on good food: on pure fragrant coffee, fresh wholesome bread, thick apricot jam, Westphalian ham, and malty beer.

The world of swindling described by Accum is, in many ways, still our world. There is the same reluctance of governments to upset the wheels of commerce, the same ability of science to invent fiddles as well as methods for exposing them, the same long and circuitous chains between consumer and producer, the same reckless willingness of the worst swindlers to sacrifice the health of others to turn a quick buck. That is why the story begins with him. Before Accum, no one had given a complete picture of how adulteration could affect every layer of society, knitting everyone together in a web of falsehood, ignorance, and poison. Then again, before Accum—and the restless, industrializing London he lived in—food swindling had not always been so bad.

2

A Jug of Wine, a Loaf of Bread

Wherefore do ye spend money for that which is not bread?
—Isaiah 55:2

What is bread? Walk into the average supermarket and you might be forgiven for thinking that bread could be almost anything that contains flour. It might be long, spongey hot dog rolls or round, dry hamburger buns. It might be cardboardy pitta or floppy tortilla "wraps"; weirdly moist "batch" bread or weirdly crusty "farmhouse" bread. It could be flavoured with any substance from dried onion to chocolate. Regardless of the colour and dimensions, this "bread" could, and probably does, include all kinds of rather unsavoury ingredients: emulsifiers; flour treatment agents; Soya flour; bleach and flavourings; hardened fat to give the crumb its requisite soft and springy texture; hidden enzymes that are not even listed on the label. Even if it has been neither proved nor fermented nor kneaded—but hastily mixed by the so-called Chorleywood method—it can still call itself "bread." More audaciously, the label will probably claim that this strange, farinaceous concoction is either "superior" or "traditional" in some nonspecified way. You may be seduced into buying it by the artificial baking smells being pumped through the store, yet no one intervenes to stop you from being deceived. No one protests that you are being sold an unacceptably high volume of water and grease along with your flour. Least of all, no one thinks of tracking down the creators of these "breads" and punishing them for their feeble products. No one forces them to eat their own loaves, no one parades them through the streets with bags of inedible sliced white hanging round their necks. What do you expect, at these prices? This is simply what we have become accustomed to.

Yet this indifference to the quality of bread would have seemed profoundly odd to our ancestors, for whom "bread" had very distinct and sturdy connotations. Bread could exist thanks only to the presence of a miller and a baker. A baker was someone with skill, whether a professional or a home baker. As the old rhyme had it:

> Blow wind, blow! and go mill, go!
> That the miller may grind his corn;
> That the baker may take it,
> And into bread make it,
> And give us our bread in the morn.

Sometimes bread was brown and sometimes white, but it always needed an exact calibration of ingredients. Governments intervened to protect the quality of bread. When bread got bad—whether through poor grain or sloppy preparation or the addition of nonpermitted ingredients such as peas and beans—customers knew it and complained. In the eighteenth century, the French police stipulated exactly how loaves should be made—to an exact weight, of nonbitter grain, "kneaded," "well risen," and "cooked appropriately."

In those days, people knew what bread was. They knew it was a staple food properly made of flour, salt, leavening (sourdough or yeast), and water—not too much of the latter. Wine was a different matter. For much of its history, "wine" could be almost anything alcoholic and grapey. Sometimes, even the grapes were missing, replaced by raisins or, in the case of some Victorian "champagnes," by gooseberries. Wine might contain honey or lead or seawater. It might be watered down or brandied up. It might be dyed with arsenic or tempered with horseradish. As Rod Phillips has said, "Wines have been heated, boiled and cooled, they have been blended, they have been mixed . . . and coloured . . . And they have all been called wine."[1] Often, it was hard to be sure whether a wine was fraudulent

A medieval woman selling bread.

or not, since the definitions of what "wine" was were not clear. People knew when wine was really bad, but they had less sense of what it would mean for wine to be really good, or simply what it meant for wine to be real in the first place.

Now, this has changed dramatically. From the early twentieth century onwards, wine quality has become infinitely better than it was in the past. Legally, wine is "the alcoholic beverage obtained from the fermentation of the juice of freshly gathered grapes." It may contain a few other things—some sulphur to preserve it and sugar added to the must to adjust the alcohol content—but these additives are properly policed by the relevant authorities. When you are buying a bottle of wine, you can be confident that it really is fermented grape juice and that it really does contain the amount of alcohol stated on the label (whether it will taste good is another matter). It has been said that "the wine commercially produced today is a far more pure and reliable product than that produced in any earlier period."[2] In the past, adulteration of wine was the rule; now it is the exception. With bread, the reverse is true, and as with wine in the past, we are no longer sure what bread is, never mind whether it has been falsified.

How did this change come about? How did bread become (generally) so bad and wine (relatively) so good? The answer to these questions tells us a great deal about the long struggle between producer and consumer to get the most out of their food and drink.

Wine

Part of the difference lies in the fact that wine is a much less natural product than bread. Whereas bread is cooked, wine is manufactured; the one is (potentially) ruined by large-scale modern industry whereas the other is (potentially) improved by it. Good wine is the result of a complex interaction between people and the environment, and for a long time people were less than expert at their side of the

Roman wine jugs.

bargain, making mistakes that they could rectify only by poisoning their own product in order to cover them up.

For all their odes to wine jars and apostrophes to "wine-dark seas," the Greeks and Romans made some badly corrupted wines. Pliny the Elder was complaining about it in the first century A.D., in his *Natural History*. "So many poisons," he wrote, "are employed to force wine to suit our taste—and we are surprised that it is not wholesome!"[3] Wine making was an imprecise art, and ancient wine-makers were much less skilled in monitoring the various stages of manufacture to achieve the taste they desired than their modern equivalents. If all the elements came together—a luscious harvest of grapes and accurate fermentation—ancient wine might well have been delicious. Often, though, wine came out wrong and needed to be "adjusted" after the event. So, in Africa, said Pliny, rough wines might be softened with gypsum and, "in some parts of the country, with lime." The Greeks on the other hand, feared blandness in wines rather then roughness and "they enliven the smoothness of their wines with potter's earth or marble dust or salt or sea-water, while in some parts of Italy they use resinous pitch for this purpose and it is the general practice both there and in the neighbouring provinces to season must with resin; in some places they use the lees of older wine or else vinegar for seasoning."[4]

Some of the additions for enhancing the wine's flavour were fairly harmless. Commonest was honey, often added in prodigious quantities—as much as half and half—suggesting that unsweetened wine must often have been a mouth-puckeringly sour drink. The Gauls apparently were fond of adding herbs, such as thyme or rosemary. The Greeks added rose petals, violets, or mint. Pliny mentions such strange additions as asparagus, rue, sorb apples, mulberries, Syrian carob pods (a bit like chocolate), juniper berries, turnips, roots of squills, cassia, cinnamon, and saffron. It is hard to know which of these were adulterations proper and which were simply culinary innovations. We do not believe we *adulterate* wine when we add cinnamon quills, cloves, orange peel, and sugar to "mull" it at Christmas time (though the quality of the wine can often do with concealing). On the other hand, some of these ancient flavourings were clearly intended to deceive. The Roman farming writer Columella recommended various artificial flavourings but advised wine-sellers not to advertise the fact "for that scares the buyer off."[5]

Other ancient wine additives had a more practical purpose. The great problem with wine was that it went off so quickly. Most Roman wine was essentially alcoholic fruit juice. There were exceptions to this—the poet Juvenal writes of vintage wine laid down in its bottle for centuries, since the time when Consuls had long hair—but these more mature wines were just this: exceptions.[6] Beaujolais Nouveau is well aged by comparison with most ancient Roman wines. We can get some idea of how quickly ancient wine deteriorated from the fact that the jurist Ulpian, writing in the third century A.D., described "old wine" as wine made the previous year.[7] Anything that could make the wine keep longer in the Mediterranean climate was seized upon, with too little thought for the effect it might have on the wine's wholesomeness. This was the main reason that so many wines, Greek wines especially, were resinated—the ancient counterparts of modern retsina. The earthenware amphorae in which wines were kept were often porous, letting air in and oxidizing the wine. Winemakers found that the wine kept better if the inside of the jar was coated with resin; and better still if a little resin was added to the wine itself along with the must, either in powdered form or as a sticky liquid. Eventually, drinkers became connoisseurs of different kinds of resins—the resin of Syria was said to resemble Attic honey—but resin's primary purpose was preservation; drinking a glass of inferior retsina can still feel a little like being pickled in Cuprinol.

Preservation was also the point of adding seawater—salt is a preservative—though it must have made for a repulsive drink. Pliny complained that wines made with seawater were "particularly injurious to the stomach, nerves and bladder."[8] Nothing like as injurious, however, as a still more popular preservative, lead, which Pliny saw as harmless; and he was not alone in this belief. The Romans thought that lead was wonderful for wine. Lead ions inhibit the growth of living organisms, so lead delayed the point at which wine turned into vinegar and generally made it less apt to spoil. What is less well known—because we do everything we can *not* to consume it now—is

51

that lead is delicious. Those Victorian children gnawing on their lead pencils and chewing on their lead soldiers were not just playing out some Freudian oral fixation, they were also satisfying a craving for sweetness. The Romans swore by lead to correct a sour wine, making it sweeter. This was especially the case if the lead was added in the form of *sapa* or *defrutum*, a concentrated grape juice or must boiled down in a lead vessel.

Writing as a landowner in the first century A.D., Columella noted that "Some allow a quarter of the must to boil away after pouring it into leaden pots, some a third," but that "if one lets half of it boil away, it becomes undeniably a better *sapa*"—and one with a higher lead content.[9] The farming writer Cato recommended using a fortieth part of this deadly reduction during wine making.[10] He would surely never have done so had he realized that lead was a poison, capable of causing headache, fatigue, and fever; sterility; loss of appetite; severe constipation and unbearable colic pains; loss of speech, deafness, blindness, paralysis, loss of control of the extremities; and eventually death. Leaded wine must have had terrible effects on the Romans—one historian has suggested that endemic lead disease may have been one reason why so many wealthy Romans were sterile—yet they continued to drink it in the belief that it was good for their health.

What made lead such a stealthy poison was that its effects are cumulative. Most of the body's absorbed lead is stored in the skeleton where it builds up over many years. Unlike food poisoning, which quickly affects all those who eat a single meal, the symptoms of lead poisoning build up variably and gradually. And so the use of lead in wine continued long after the Romans, into modern times, reflecting the fact that without lead, or some other preservative, "most wine was so unstable that it was likely to deteriorate and go bad within a year of being made, even when it was not shipped over long distances and exposed to rough handling and fluctuating temperatures."[11] Sometimes there would be a mini-epidemic of wine-induced lead poisoning, especially after a cold summer when the sourness of the grapes encouraged wine-makers to overdo the lead. This disease came to be known as the colic of Poitou or *colica Pictonum*, because the citizens of that French town had suffered an especially bad epidemic. Still the connection wasn't made between the colic symptoms—of severe abdominal pain, weakness, and nausea—and the lead. Doctors were more apt to blame the sourness of the wine than the lead that had been used to counteract the sourness.

It was only late in the day, at the end of the seventeenth century, that the danger of lead in wine was spotted, by a German medic called Eberhard Gockel, city physician of Ulm in the Baden-Württemberg region. At that time, lead was still a common addition to wine, though no longer as *sapa*. Now, the lead additive usually took the form of litharge—a foam or "spume" produced during the refining of lead—or sometimes *Bleiweiss* (lead oxide) or ceruse (lead carbonate). How did wine-makers get away with these deadly additions? There had been plenty of laws against the adulteration of wine. In 1487, the city of Ulm had passed a law requiring every innkeeper to swear that his wines were pure and that neither he nor his wife nor his servants had added any of a list of additives, which included *Bleiweiss*. Ten years later, an imperial edict forbade various wine additives, including ceruse. Yet these laws were widely flouted. Penalties for doctoring wines were "surprisingly lenient."[12] In the fifteenth century, German wine adulterers were punished by money fines plus public shaming, and their wine was poured into the river. This was partly because legislators did not realize the effects of lead on the human body; once they did, several German cities issued laws specifically forbidding the use of litharge in wine with stringent penalties—imprisonment and death—for those who broke the law.

It was by accident that Gockel discovered the evils in the habit of "correcting" wines with lead. Over the Christmas holidays in 1694—a time when festive cheer may have encouraged some of the monks to indulge a little too heartily—the prelate of the Teutonic Order of the Knights of Ulm and several of his monks developed colic. Gockel was summoned. His first impulse was to check the wells and the kitchen, neither of which had anything amiss. Gockel noticed that all the monks who had not drunk wine had been spared the colic. Then, on one of his visits to the ailing monks, he accepted a glass of wine and soon found himself feverish and with chronic colic pain.

After much investigation, he eventually tracked down the recipe for this wine from a dealer near Göppingen and found that it contained litharge. At once, lead became the prime suspect. To make sure, Gockel did some experiments. He took the "worst and sourest wine" he could find and by adding litharge managed to convert it into the "best and loveliest of wine"—the word he used was *süsselectenliebli-chen*, one of those untranslatable German words, conveying a zenith of delectable sweetness.[13] Loveliest except for the fact that it caused horrific stomachaches.

Gockel's discovery did not entirely eliminate lead from alcoholic drinks, however. In the late eighteenth century, there was wide incidence in the English county of Devon of "Devonshire colic"—the result of lead poisoning from cider. Opinion differed as to how the lead got there. In 1767 and 1768, George Baker published articles arguing that the cider was contaminated by lead in cider presses. In 1778, however, James Hardy insisted that it was not the presses that were to blame but the lead-glazed earthenware from which the poor of Devon drank their hard apple cider. The acid in the cider dissolved some of the lead in the glaze. This explained why the poor were the most susceptible to Devonshire colic, since they were the ones who used earthenware, while the rich drank out of safer vessels made of glass and stone.

This lead poisoning was accidental. Similar problems arose when wine bottles were cleaned with lead shot. But the deliberate addition of lead to wine carried on too. In 1750, the tax inspectors of France were "astonished" by the huge volume of spoiled wine—*vin gâté*—being brought into Paris; this inferior wine was used, perfectly legally, to make vinegar. In this case, however, a number of wine merchants had registered themselves as vinegar merchants in order to have access to this cheap vinegary wine, which they touched up with litharge and then sold, presumably at a great profit, as real wine.[14] Still more blatantly, a book of 1795 called *Valuable Secrets Concerning the Arts and Trades* recommended sweetening "turned wine"—in other words, wine that had turned acid—with a "quarter of a pint of good wine vinegar saturated with Litharge."[15] It makes you wonder how such attitudes could persist, so long after lead had been proved a poison. Frederick Accum answered this question in his *Treatise on Adulterations* of 1820. Lead continued to be added to wine, he argued, because "there appears no other method known of rapidly recovering ropy wines." As for the moral dimension, "Wine merchants persuade themselves that the minute quantity of lead employed for that purpose is perfectly harmless and that no atom of lead remains in the wine. Chemical analysis proves the contrary; and the practice of clarifying spoiled white wines by means of lead, must be pronounced as highly deleterious." To Accum, the vintner or wine dealer who "practises this dangerous sophistication, adds the crime of murder to that of fraud, and deliberately scatters the seeds of disease and death among those consumers who contribute to his emolument."[16]

Accum's publicity probably contributed to the decline in the use of lead in wine. It was also harder to doctor wines with lead once reliable chemical tests were developed for its detection. In 1818, the French scientist Mathieu Joseph Bonaventure Orfila (1787–1853) listed no fewer than nine different tests for wines adulterated with lead; for example "they scarcely redden the tincture of turnsole because the acid which they naturally contain is saturated by the oxide of lead."[17] But still the dosing and doctoring of wine continued, in manifold forms. Wine making was an art that liked to shroud itself in mystery, and the trade secrets being hidden were often criminal. A Renaissance critic complained that the vintner was "a kind of Negromancer, for at midnight when all men are in bed, then he . . . falls to his charms and spells, so that he tumbles one hogshead into another and can make a cup of Claret that has lost its colour to look high with a dash of red wine at his pleasure."[18] In strikingly similar terms, more than a century on, in the *Tatler* magazine of February 1710, Addison observed that

> There is, in this city, a certain fraternity of chemical operators, who work underground in holes, caverns and dark retirements, to conceal their mysteries from the eyes and observation of mankind. These subterraneous philosophers are daily employed in the transmutation of liquors, and by the power of magical drugs and incantations, raising

under the streets of London the choicest products of the hills and val-
leys of France. They can squeeze Bordeaux out of the sloe, and draw
Champagne from an apple.[19]

These "subterranean philosophers," also known as "wine brewers"
because they could brew up wine from virtually anything, had their
counterparts all over Europe. Sicilian growers in the nineteenth century were notorious for adding chalk to many wines, not only to
preserve them, but to speed up clarification and improve the wine's
colour.[20] This chalked wine—*vino gessato*—became so common that
foreign buyers lost trust in Sicilian wine and were made to pay a premium for nonchalked wine—in other words, forced to pay extra for
something that should have been standard all along.

French wine was regularly falsified too. During the 1870s, when
French vineyards were struck by vine blight, there was a "notable
increase in dishonesty."[21] To maximize yields, wine-makers started
to make watery brews from the second, third, and even fourth press-
ings of grapes. Then, to give an illusion of body, they would be coloured with fuchsine, which contains arsenic. Things hadn't really
progressed much from the days of Chaucer, whose Pardoner urged:

> Keep clear of wine, I tell you, white or red,
> Especially Spanish wines which they provide
> And have on sale in Fish Street and Cheapside.
> That wine mysteriously finds its way
> To mix itself with others—shall we say
> Spontaneously?—that grow in neighbouring regions.[22]

Until the 1900s, when a much stricter classification system came in
for French wine and the wine-making industry modernized its techniques, the story of wine swindling is one more of continuity than
of change. It assumed many forms, but the universal trick was the
attempt to pass bad wine off as good. In Bordeaux in the 1850s, they
might do this through chemical wizardry, distilling the bouquet of
fine claret from various artificial potions.[23] In the fifteenth century,
the chemicals were more basic—one source mentions "eggs, alum,

gums and other horrible and unwholesome things," but the fundamental aim was the same: to touch up undrinkable wine.[24]

How was such dishonesty policed? With endless, and endlessly evaded, punishments and prohibitions. There have been laws against wine adulteration ever since Charlemagne, who, in 802, issued what was probably the earliest edict against fraudulent wine in the post-classical age.[25] Indeed, it has been said that "for there to be a wine fraud" in the first place, "there must be a wine law."[26] The fourteenth and fifteenth centuries were particularly rich in wine laws, many of them local to certain cities. In 1364, John Penrose, a vintner, was called before the mayor of London and found guilty of selling bad wine. He was ordered "to drink a draught" of his own bad wine and have the remainder poured over his head, before abandoning "the calling of vintner in the city for ever";[27] if he sold bad wine elsewhere, though, the mayor wasn't going to stop him. Also in 1364, the town council of Colmar in Alsace made it an offence for an innkeeper to adulterate wine by adding water, brandy, sulphur, salt, or any other ingredient.[28] There were also countless attempts by the authorities to stop vintners and innkeepers from mingling different kinds of wine, and/or passing off plonk as fine wine. In 1419, a "William Horold" was convicted of "counterfeiting good and true Romney wine. He took old and feeble Spanish wine" and added "powder of bay" and other "unwholesome" powders.[29] Similarly, in 1415 the town innkeepers of Bordeaux were summoned to the town hall and told that if there were any more cases of passing off other wines as those of Bordeaux, the offenders would be put in the pillory and banished from the town.[30]

In England, the fourteenth century saw detailed legislation to protect wine drinkers from fraud. On 8 November 1327, Edward III decreed that weak and out-of-condition wine could not be mixed with any other.[31] By the same law, he determined as well that every customer had the right to see his wine being drawn from the cask, and it was forbidden to put a curtain over the doorway leading down to the innkeeper's cellar. New and old wine were not be blended, or even stored in the same inn. Rhine wines could not be sold by someone who sold the wines of Gascony, La Rochelle, and Spain.

This was a remarkably explicit decree. So why didn't it work? Why was Accum in 1820 still complaining about "factitious" wines—of "port" flavoured with a tincture of raisins; of wine corks dyed red to look as if they had been in long contact with the wine; of weak wines given "a rough austere taste, a fine colour and a peculiar flavour" with the use of astringents?[32] One of the reasons wine was especially prone to adulteration was that, unlike ale, it was a luxury good, and an imported one. Premium comestibles—tea in the nineteenth century, spices always—are especially tempting for fraudsters. Why therefore did wine not lose its reputation in the centuries before the twentieth century, when it was such an unreliable drink? Why, despite all the frauds to which it was subject, did it retain a glamour and even an image of wholesomeness?

Partly, this must have been because not all wine was debased. In a good year, when the harvest was full and ripe, the wine-makers of Burgundy or Chianti must have made some fine bottles. Even if they did not understand the science of what they were doing, hundreds of years of experience must have given wine-making families the ability to make enjoyable and wholesome wines from good grapes. The problems mainly arose, as we have seen, when the grapes themselves were not good, which was when even respectable vintners resorted to chemical adjustments.

Sometimes, though, wine's reputation was simply relative. During the eighteenth century, wine could seem healthy because at least it wasn't gin, a drink for which there was an unstoppable craze in the first half of the century. In 1726, there were 6,287 places in London where gin was sold, much of it rasping with turpentine or sulphuric acid. If the adulteration of wine was common, the adulteration of spirits was commoner. In the history of distilling, adulteration is the rule, not the exception. The trade had always been rife with diluters, "artificial Rectifiers," and "sophisticators," who would draw brandy from a turnip, or "meliorate" spirits with green vitriol.[33] During the gin craze, the desire of the poor for penny drams of hooch drove both licit and black market distillers to fabricate "gin" from whatever grain and flavourings they could get. Wine was only sometimes poisonous. Even when pure, gin was a kind of poison, "Mother's Ruin" as the anti-gin campaigners insisted—a drug that got into the milk of babies through their drunken nurses, or turned women to depravity and men to disorder.[34] By the Gin Act of 1736, the British state even attempted—unsuccessfully—to ban gin. By contrast with this juniper-sodden firewater, wine seemed a moderate drink. In *The Wealth of Nations*, Adam Smith looked enviously to the wine-drinking nations of Europe as being soberer than Britain.

Many wise voices over the ages told the English that they would be better off drinking fruit wines made from native British fruits than fancy—and often faked—foreign wines. Sir Hugh Platt, an Elizabethan courtier, beseeched his contemporaries to take an "English and naturall drinke" made from Royston grapes instead of poisoning themselves with imported "concocted" wines.[35] "Native" fruit wines were one of Accum's passions, too. Yet this could not be the whole answer, because swindled wines were for sale too in France and Italy, where they were native. The truth is that until the process of wine making was itself more reliable, there would always be a large volume of ropy wines; and ropy wines led inexorably to fraud. Legislation could try to protect the consumer and the revenue from fraud, but the chances of success were not good, until reliable wine making was the norm.

"The best test against adulterated wine is a perfect acquaintance with that which is good," wrote the wine expert Cyrus Redding in 1833.[36] He was thinking of British consumers, whose palates were often so numbed by alcohol that they could not tell when their port was "cut" with spirits or when their claret was fake. Merchants in Bordeaux knew that when they were selling to the English market, they could get away with mixing their wines with 10 percent benicolo, a cheap, alcoholic Spanish plonk, because English wine drinkers didn't know the difference.[37] Yet, increasingly, the French consumer was not much better off. Things had to get very bad in French wine production, in the nineteenth century, before they could get better. There was an explosion of new substances that became a part of wine making: sulphuric acid to give wines a little sharpness, "allume" added to fix the colour in falsely coloured wines, salicylic acid to stop fermentation, iron sulphate to even out the taste.[38] Suddenly, it was hard to know on what basis to judge wine quality. Previously, taste had seemed enough, but these new chemical substances "rendered these traditional criteria obsolete."[39] Then, in the 1880s, phylloxera hit, the tiny root-feeding aphid that wiped out almost 2.5 million hectares of vineyard in France. Vines were also affected by mildew. Phylloxera, it has been said, led to "a notable increase in dishonesty," especially in the mid-1880s.[40] The shortage of real grapes meant that vignerons were forced to desperate measures. Vast amounts of raisins were imported from Greece to Marseilles to fabricate raisin wines, labelled as the real thing. And "wines" were even sold that had no vine product in them at all, just chemicals, sugar, and water.

Both wine-makers and the French state recognized a desperate need to set new norms for wine making, to find new definitions of what "wine" actually was. Special new laws were passed dealing with adulteration. In 1889, raisin wines were specifically outlawed; in 1891, the practice of "chalking" was prohibited; in 1894, it was forbidden to sell either watered-down wines or wines laced with extra alcohol.[41] In the meantime, Louis Pasteur had began to establish the science that would finally enable reliable avoidance of some of the most common failings in wine, without recourse to swindling. In the 1860s, Pasteur identified many of the microorganisms that caused different faults in the bottle. An excess bitterness was due to degraded glycerol; flabbiness was caused by a polysaccharide. In Pasteur's view, "yeast makes wine, bacteria destroy it." This was the foundation of modern oenology. It would take several decades before Pasteur's bacteriology was fully applied in the vineyard, but at last there was now the means available for scientific wine making.

In 1905, the French government issued a new law on wine quality, defining wine as the product of "fresh grapes." This was a start, but many fraudulent wines were still being marketed, and there were numerous examples of generic wine being labelled as if it came from one of the prestigious wine regions such as Burgundy. Ultimately, it was not prohibition by itself but enhanced standards of wine making that resolved the problem. The Appellation Contrôlée system, which evolved in the 1920s with Châteauneuf-du-Pape but was officially established in 1935, laid out extremely detailed rules for every stage

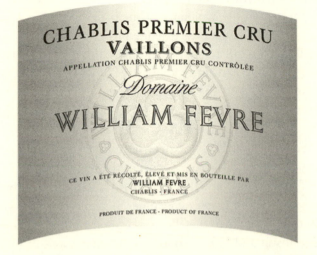

CHABLIS PREMIER CRU
VAILLONS
APPELLATION CHABLIS PREMIER CRU CONTRÔLÉE
Domaine
WILLIAM FEVRE

CE VIN A ÉTÉ RÉCOLTÉ, ÉLEVÉ ET MIS EN BOUTEILLE PAR
WILLIAM FEVRE
CHABLIS · FRANCE

PRODUIT DE FRANCE · PRODUCT OF FRANCE

Wine bearing the "AOC" mark on its label, a guarantee that the wine has been produced under strict conditions of quality control and within a defined geographical area.

of the wine-making process, drawing on new expertise in oenology. Appellation Contrôlée is fundamentally a system of geographical control, according to which "Bordeaux," for example, must come from the region of that name and nowhere else. But AC also set rules about vine varieties, pruning and training methods, alcoholic strength, and the quality of the grapes used.

The system was not without teething troubles. Attempts to define which region was allowed to produce which wine could lead to bitter disagreement, as in Champagne, where there were riots over the question of whether the nearby Aube region was allowed to produce true "champagne" (in 1927 it was decided that it was). But for all its rigidity and arbitrariness, most wine-makers found the new system attractive, since it protected them from unwanted imitators and kept market prices high. The system required higher-level policing than that required by a simple ban on adulteration—the French Service de la Répression des Fraudes still does spot checks to make sure that a Médoc really is what it says, and not some clever imitation. For consumers, the AC and AOC marks—which have now been extended to

61

foods as well as wine—were, and are, a guarantee of quality. They are based not just on the avoidance of the bad, but on knowledge of the good—a knowledge that has extended from producers to consumers.

Hardly anyone disputes that the past three decades have seen a transformation in consumer knowledge of wine. In 1976, when wine critic Robert Parker, then a young lawyer, established his famous 100-point system, wine drinking was a minority interest, an arcane pursuit, in both Britain and the United States. By 2006, new research was predicting that U.S. wine consumption would overtake that of the French within three years. Most of the wine being drunk is modest stuff, which may not rank highly on a Parker scale, but there is, nevertheless, much greater consumer awareness of how good wines should taste. Thanks to the efforts of Parker in the United States and Hugh Johnson and Jancis Robinson in Britain, and of films such as *Sideways*, you no longer have to be a "wine buff" to know that sauvignon blanc might taste of gooseberries or that shiraz (or syrah) has overtones of chocolate. This knowledge might seem useless, but it is actually very powerful. Jancis Robinson argues that "passing off has become increasingly difficult and, just possibly, less rewarding as wine consumers become ever more sophisticated and more concerned with inherent wine quality."[42] In the 1950s, there were cases of "fine" wine merchants buying a single vat of wine and passing it off as Beaune or Beaujolais or Burgundy, depending on what the customer asked for.[43] It is much harder to imagine anyone getting away with this now. Cyrus Redding was right. "The best test against adulterated wine is a perfect acquaintance with that which is good"; and this acquaintance is far more widespread now than at any time in the past.

It would be naive, though, to suppose that wine adulteration is entirely a thing of the past. Many are concerned by the rising alcohol levels in red wine, from around 12 percent in the early 1980s to around 14 percent now, a result of the vogue for very sticky reds, particularly in the New World. Such wines can make drinkers far drunker far quicker than they had intended. On the other hand, at least the alcohol content is stated on the label.

Periodic scandals still rock the wine world. In 2005, a South African producer, KWV, was found to have added artificial flavouring to two batches of sauvignon blanc.[44] In 2006, the best-known producer of Beaujolais, George Duboeuf, was charged with tweaking some of his wines with non-Beaujolais grapes, a charge Duboeuf appealed against.[45] More serious was the Austrian antifreeze scandal of 1985, when a hundred Austrian wine-makers were indicted with boosting the body and sweetness of their wines with diethylene glycol, a chemical used in antifreeze.[46] The next year came the Italian methanol affair, which killed more than twenty people: gallons of cheap Italian wines were found to be adulterated with methyl alcohol, a toxic version of alcohol that can cause blindness or death. Yet the net result for wine quality of this skulduggery was probably beneficial: a new artisanal wine industry sprung up in Italy to eradicate the memory of the scandal. In the past, one wine scandal was likely to lead to another, putting ideas into the heads of those who might not otherwise have contemplated the fraud. Now, one wine scandal is likely to lead to better enforcement of standards. On balance, wine is purer, more honestly labelled, and—possibly—more delicious than at any time in the past.

All this improvement has happened in the last hundred years, a mere fraction of the total history of wine consumption. During the long period when there were no reliable tests for the quality of wine, drinkers keen to assert their rights tended to fall back on the business of quantity instead. Alongside the enduring and largely futile search for better wine, therefore, went another, apparently contradictory quest: for accurate and honest measures, recalling the Woody Allen joke from *Annie Hall*: "The food here is terrible." "Yes, and such small portions." The wine was often terrible and sometimes noxious. And still consumers wanted to make sure they were getting their full measure of it.

Weights and Measures

Being short-changed is unquestionably a nasty feeling. It makes you feel foolish, to be given a pint of beer that is all head and no beer, or to be charged for a kilo of plums when all you have bought is a pound, or to leave a shop and find that you are a lemon short or that you've been slipped a mouldy punnet of berries in place of the fresh ones you saw on display. You are left simmering with a mixture of embarrassment and rage. Do you return to the shop and confront the seller, giving yourself an inconvenient journey and risking an aggressive dispute at the end of it? Or do you let your resentment simmer away, because you feel powerless to do anything about it? This dilemma goes back a long way. Giving short weight is surely one of the oldest of all the food and drink swindles. Traditionally, or so the joke goes, merchants had three sets of weights: a heavy one for buying, so they got more than their fair share; a light one for selling, so they gave less than they should; and an accurate set for themselves, so that they knew what things really weighed. The potential dishonesty of sellers has always been one of the earliest concerns of the law.

Magna Carta, the peace treaty signed by King John and his rebellious barons at Runnymede in 1215, is now seen mainly as a noble and lofty document about liberty. It established the principle of habeas corpus, protecting the citizen against wrongful imprisonment, and has been revered through the ages as the cornerstone of constitutional justice. But in truth it was a ragbag of a charter, the work of many hands and minds, a solution to diverse grievances, which settled none of them adequately. And if it was concerned with liberty, it was no less concerned with fair measures for ale and wine. Article 35 reads: "There shall be one measure of wine throughout all our kingdom, and one measure of ale, and one measure of corn, namely the quarter of London."

Behind this bland statement lay a politically fraught but hugely potent alliance. On the one hand, there were the consumers, fighting an often losing battle to be served an honest measure of food and drink. On the other, there was the British crown, seeking a standardized measure as a way of spreading the rule of law, regularizing commerce and getting the full amount of any customs or sales tax due to the Exchequer. Establishing agreed weights and measures was one of the first aims of good governance; but it was easier said than done.

William the Conqueror had commanded that all weights and measures be uniform and stamped with his seal to authenticate them. The weights and measures he chose were Anglo-Saxon ones, the standards for which William installed first in Winchester and then in the crypt at Westminster Abbey as a sign of his control over his new kingdom. Despite this, there was huge regional variation in the weights and measures used. "A pound" of corn in York might be very different from "a pound" of corn in Hastings. This was partly because—despite the crown's attempts to unify—some localities clung to regional standards of measurement. Back in Celtic times, each tribe had had its own system of units tailored to the particular needs of the group; in places, this practice survived. Measures of ale might vary even from one hamlet to the next. Oversight was haphazard to say the least, and it has been reported that "the alewife who was thought to have given short measure would soon be the alewife floating in the village pond."[47]

Even national standards, insofar as they existed, fluctuated. Units of weight were determined essentially by the weights of grain and money. The basic unit was the grain of barley—or later, the grain of wheat, which is marginally lighter. The old Etruscan pound used by the Romans was 4,210 grains, but as new coinage was introduced, the pound rose in weight. The Saxon king Offa (r. 757–96) brought in a new "moneyer's pound" to correspond with his new silver penny, the *sterlingus*. Offa wanted twenty of these pennies to designate an ounce (450 grains) and twelve ounces to a pound (5,400 grains). Over time, different weights were set up for measuring different things. The *wool pound* of 6,992 grains was heavier than the *tower pound* of 5,400 grains used for measuring coins and the *troy pound* of 5,760 grains, used to measure gunpowder. Today we use the *avoirdupois pound*, the word deriving from the French for "goods of eight," equivalent to 7,000 grains.

The situation was further complicated by the fact that most of the countries that Britain did trade with—Scotland, France, Flanders, the German Hansa, Italy—had their own weights and measures, which did not correspond with the British ones. The North German commercial pound corresponded to 6,750 grains of wheat, while the old Etruscan pound was 4,210 grains and the Saxon pound was 7,680 grains. The crown's difficulty in cutting through this confusion is shown by Richard I's Assize of Measures of 1197. It called for uniform measures throughout England but did not say what these measures should be. Magna Carta did at least state that the measurement for corn should be the London quarter—equivalent to eight bushels. But it named no measures for wine and ale.

As the Middle Ages progressed, the crown's attempt at standardizing weights and measures became more effective. In 1266, exact weights were set for both bread and wine, and it was also decreed that an English penny "round and without any clipping, shall weigh 32 wheat corns." The crown had a keen interest in fighting fraud. In 1380, new regulations were set for wine measures imported into England, Wales, and Ireland. Parliament ordered that the king's "gaugers" carefully check the measurements of all incoming casks and punish those who refused to surrender them.[48] In 1389, Richard II decreed that all illegal weights and measures be burned. This was to counter the problem of merchants "hastily repairing illegal weights and measures to pass government inspections and then afterwards counterfeiting them to suit local preferences."[49] All kinds of sharp practice evidently went on, such as installing a larger than statutory measure with a false bottom, which would be removed after it had been inspected. Even when weights had not been tampered with, standard weights and measures often got worn down through wear and tear and were therefore far less accurate than modern platinum and gold standards. If wooden, standards became worm-eaten. If made of lead, iron, or brass they would oxidize. They were also continually expanding and contracting in the weather. A brass yard-bar was shorter in winter than summer because the metal expanded in the heat.[50]

Yet for all these problems, the question of weights and measures—unlike the question of the quality of wine—was at least something that the state understood. Policing weights and measures was one of the fundamental duties of all town officials.[51] The crown—or Parliament—did its best to set standards. These were then distributed to all the shires and from there to all the towns, where the physical measures were often attached to the walls of local municipal buildings. Finally (as of 1266), six "lawful" men went out to gather the weights and measures of individual businessmen in the town and test them for any variance from the crown standard. It was law that each local weight or measure had to be inscribed in legible script with its owner's name.[52] Punishment for false weight was usually the pillory.

This policing was also extended to particular foodstuffs, chief among them being bread. There is an old saying that if you put a baker on the scales after he dies, he will be found to be of short weight. Nevertheless, governments have always tried to prevent bread being sold under weight or at unfair prices. One of the primary functions of the British state—and in this respect it resembled other states—was to set the weight of bread and, up to a point, describe its composition. Unlike wine, whose quality was, as we have seen, largely a mystery until the twentieth century, bread was a substance whose quality could be clearly defined and gauged. The state therefore did not have to confine itself to questions of quantity when it came to bread; it could deal in quality too. The whole community knew what good bread was. It was the duty of the state to ensure that they got it.

The Policing of Bread

Fifty years after his accession, in 1266, King Henry III promulgated his famous Assize of Bread and Ale. Though it would be tinkered with over the years, the assize remained a part of British law for over six centuries, only finally being abolished in London in 1822, two years after Accum's *Treatise on Adulterations* appeared.[53] The assize regulated the weight, price, and composition of bread to a remarkably precise degree. It stipulated the size and price of a loaf of bread, depending on fluctuations in the grain market. For example, when a quarter of wheat sold for twelve pence, a standard wastel loaf (costing a farthing) must weigh 6.8 troy pounds. When a quarter of wheat sold for three shillings, the weight of a wastel loaf went down to 2.4 troy pounds. This was to stop bakers from profiteering. For most of history, baking has been seen not just as a trade but as a public duty. Bakers were expected to extract a certain number of loaves from a certain weight of flour. They could not make up recipes according to whim but had to tailor them precisely to the economic needs of the community.

A medieval manuscript referring to the Assize of Bread.

The 1266 assize mentions seven different kinds of loaf. *Wastel* bread was the standard superior loaf, which varied in weight depending on the price of corn. It was white and eaten by the rich. *Cocket* bread was like wastel, but made of slightly inferior flour. *Simnel* bread was a rich cakey thing, also eaten by the wealthy. Bread of whole wheat (*panis integer*) was the bread eaten by the masses, unless they were really poor, in which case they might eat *treet* or "household bread" made of coarse brown unbolted meal; poorer still, and you would have to torture your stomach with something called "bread of common wheat" or "horse bread," a loaf made from the miller's refuse, which cost half as much as the cocket bread.

Horse bread may not have been very nice, but at least those eating it were not being swindled. The point of the law was precisely to ensure that no deception took place. Both millers and bakers were strictly regulated. If you were a maker of coarse bread in London, you were not allowed to sell any fine bread, the obvious implication being that there was a danger of being slipped a coarse loaf for a fancy one. At Ipswich, similarly, there were four categories of bakers. One lot (the most skilful, it seems) were allowed to bake best cocket, wastel, and treet; another lot, only simnel and treet; a third group, whole wheat and second-class cocket; and the lowliest bakers could make only whole wheat and common wheat bread. Thus, the consumer, whether rich or poor, was protected.

The modern supermarket loaf is almost completely anonymous. To whom do you complain, if you are sold a bad loaf—one that is stale or underbaked or corrupted with foreign bodies? Not to the person who baked it. Their role has been reduced to little more than that of a machine operator. Instead, the person ostensibly "responsible" will be some customer care manager in an office untouched by yeast and flour. In this way, the aggrieved customer never really sees his or her score settled. You might be given some token monetary compensation, but you are never truly able to say you have held to account the person who made the thing you put into your mouth. Effectively, this is food with no person behind it. By contrast, bread in the Middle Ages was personal. Bakers were obliged to indent the bread with their seal, so that if they did break the assize, it would be easy to track them down and hold them accountable. This seal could be a mark of pride, if the bread was especially good, or of shame, if it was tampered with, or of short weight, or made with the wrong flour. Bakers were obliged to sell bread by

69

their own hand, or that of their servants, and were not allowed to use middlemen.[54]

Such measures acted as strong disincentives to malpractice. In addition, at least four times a year, bakers and millers were inspected by bailiffs, mayors, or other town officials, who checked that the assize was being observed. Despite this policing, there were always a few bakers who broke the assize. A sixteenth-century writer railed against them, saying that "the poor cry out" and "the rich find fault" with bread, while "the Lord Mayor and the Sheriffs . . . every day walk abroad and weigh your bread, and yet all will not serve to make you honest men."[55] Local court records show numerous cases of bakers selling defective loaves, though almost always of short weight rather than badly composed.

Yet if the law did not ensure universal honesty, it did mete out regular punishment when dishonesty was detected. London bakers seem to have been singled out for retribution. The first time a baker was caught swindling, he was drawn from a hurdle from the Guildhall, through the dirtiest streets, with the bad loaf hanging from his neck. The second time, he was drawn from the Guildhall through Cheap Street, with his head and hands locked through a wooden pillory, for all to ridicule. The third time, the poor wretch was dragged on a hurdle, his oven was dismantled, and he was forced to swear that he would never again practise the trade of baking in the City of London.[56] In later centuries and in other towns, the punishments were often more lenient. In Southampton in the fifteenth century, the norm seems to have been confiscation of the offending bread, plus a fine, often quite a modest one.

Though the punishments varied from place to place and time to time, what remained universal was that people knew the composition of bread, in a way that few people do nowadays. Some might say this is a sign of our luck. We can afford not to scrutinize the dubious contents of a sliced loaf, because we have plenty of other things to eat. By contrast, our ancestors had no choice but to know the composition of bread since they depended on it for their subsistence. This explains the widespread mistrust of miller and bakers: they were suspected because people depended on them to fill their bellies. Baker's bread was relatively expensive, and the majority of poorer households avoided the potential swindling of bakers by kneading their own dough at home, forming it into loaves and baking it in communal town ovens, or else in a bakeshop. But they still depended on the miller, who was often seen as a malign figure hoarding more than his share of the grain that he was meant to grind for the whole community. As the miller says in the old song "The Miller of Dee," "I care for nobody, no, not I / If nobody cares for me."

In prerevolutionary France, similarly, millers and bakers were demonized—often unfairly—precisely because their job was so important. "To eighteenth-century Parisians the presence of bad bread in the markets or shops was an intolerable affront and menace. It signalled either an act of social crime or a mark of social breakdown"[57]— social breakdown, because in eighteenth-century Paris, as in medieval Britain, the baking of bread was highly controlled, and bad bread was a sign that basic law and order had failed. As in Britain, the greatest problem was short weight. Stephen Laurence Kaplan, the great historian of French bread, calculated that the bread lost to consumers through short weight in Paris alone would have been enough to feed "several thousand families."[58]

The response to this problem was a draconian policing system. The philosopher Diderot told a story of a Turkish judge or *kadi* who described what happened when he heard of a baker who had sold a short-weight loaf. "I went to his bakery. I had his bread weighed and found it light. His oven was still red hot. I had him thrown in, and my business was finished."[59] Parisian justice for bakers was not so summary, though it could still be severe. If consumers thought they had a short-weight loaf, they could rush it to a police commissioner, who would then verify it, and, depending on what was found, summon the baker and fine him. Should he fail to pay at once, he might be jailed immediately. His stove might be dismantled, and his shop walled up for up to two years, or if he sold in a market, he could lose his market stall, and hence his livelihood. In the general run of things, bakers were often intimidated and taunted by the police, who knew they held their reputations in their hands.

71

A dealer in fraudulent goods at a medieval Islamic marketplace being punished and publicly shamed. From a seventeenth-century Persian manuscript.

Unsurprisingly, many bakers found this system unfair. By law, they were obliged to mark each loaf with its weight. What they were not allowed to do, however, was to use scales to weigh the bread at the point of sale. An amount of guesswork was therefore involved in constructing loaves to the right size, since a piece of dough going into the oven does not weigh the same as the bread coming out. Bread loses weight as water evaporates. To the bakers, it seemed crazy to expect exact weights under these circumstances. In 1743, the bakers of Paris addressed a collective petition to the chief of police pointing out what a very variable process baking was. The existing law seemed to expect them to control the "four sovereign elements" of nature—earth (i.e., flour), air (i.e., the fermenting and proving stage), water, and fire (i.e., the oven's heat, which could never be depended on to remain absolutely constant). This was impossible. The bakers begged to be allowed to use scales in their shops, and give prices based on that weight, rather than be exposed to "calumnies" and "economic ruin" under the current system. Yet it would be another hundred years before their plea was answered—only in 1840 did a police ordinance order the sale of bread by weighing at the point of sale.[60] Until that time, the bakers managed as best they could to avoid ruin.

The chemist Parmentier, the same Parmentier whose essay on chocolate Accum had stolen from the library of the Royal Institution in 1820, took the bakers' side against the police, insisting that when it came to bread, a "shortage" of weight did not necessarily mean that swindling had taken place. There were, he insisted, a "multitude of accidents that cause weight to vary infinitely from place to place and moment to moment." Even a scientist such as himself could not ensure that a baked loaf came out at the same weight every time. He observed that prudent bakers, anxious not to be ruined, had been forced to add a "bonus" of extra dough to their loaves, ten ounces extra for a four-pound loaf, double that for a twelve-pound loaf. Moreover, the obsession with the weight of bread could actually damage its quality, since an underbaked, soggy loaf weighed more than a perfectly baked, crusty one.

Thus the fight against swindlers could end by defeating the very purpose it was designed for. There is no question that the general populace was paranoid where bakers were concerned. During the revolution, a Paris crowd threatened to hang a baker who had been denounced for selling at short weight. The people were harder on bakers than on other food-sellers because, while butter and cheese and wine might mean pleasure, bread meant life. Every ounce mattered. Thus while a large degree of wine adulteration was tolerated, the slightest tampering with a loaf of bread was a sign of doom. It was one thing to give short weight; but when the basic ingredients of bread were tampered with, this signalled not just disorder but famine.

Famine Foods

As we have seen, the bread sold by bakers before the industrial revolution was relatively pure. There were always rumours of bread mixed with ashes or sand—and genuine cases of gritty flour, if it had been badly milled—as well as the odd case of bakers pulverizing stale bread and kneading it into fresh dough;[61] but generally speaking, in good times, bakers' bread could be relied upon to be nothing but flour, leaven (whether yeast or sourdough fermentation), salt, and water. At various times in history, however, when food was short, rather than being adulterated by the seller, bread might be adulterated by the consumer for his or her own consumption. Such bread is known as famine bread or surrogate bread.

Since ancient times, famine has led to the desperate consumption of unfamiliar foods. The pattern was generally as follows. First, peasants would eat livestock that were not ordinarily meant to be slaughtered—asses, donkeys, and so on. Then, they would move on to damaged or poor-quality cereals (such as sprouting or rotten grain, the sort of thing that induces nausea). If the famine continued, they would be reduced to chewing on animal food, such as acorns or vetch. Then came the last resort—the last resort, that is, before cannibalism: the consumption of natural products that were not really foods at all, such as leather, tree bark, twigs, inedible leaves. Galen writes of the countryfolk of Asia Minor being forced to eat "twigs and shoots of trees and bushes and bulbs and roots of indigestible plants."[62]

In times of less extreme hardship, the peasantry of Britain were forced to create "breads" from all kinds of substances that would not usually have been put in a dough, such as peas and beans, rice and millet. In 1596, Sir Hugh Platt brought out a book of advice entitled *Sundrie New and Artificiall Remedies against Famine*. The first remedy, insisted Sir Hugh, was prayer. But if that failed—and evidently, it often did—there were all kinds of little tricks to stave off hunger. Sir Hugh recommends making a cheap and savoury bread of pompions (pumpkins) and fashioning sweet cakes out of parsnips. Licorice could be chewed to "satisfy thirst and hunger." He also mentioned various ancient techniques, including a weird-sounding bread made of the powdered leaves of pear trees.[63]

For the starving, there is a constant cost–benefit analysis going on in the brain. How low will you stoop to fill the gnawing hole in your stomach? How do you weigh hunger against disgust? How many inedible ingredients do you have to combine before it stops being "bread"? How much strange fodder can you eat before you stop being human and revert to the condition of an animal?

Sometimes, eating famine foods led to total loss of mind. Piero Camporesi has written of the "collective stupefaction" among medieval Italian peasants resulting from eating famine foods, a kind of "narcosis induced by adulterated bread."[64] The wild herbs used to pad out grain sometimes had drugging effects. Bread made from darnel, for example, induced a strange drunkenness in those who ate it, a dazed state in which people became either intoxicated or desperate, banging their heads against walls.

Famine "breads" have been made, and still are made, wherever in the world there is famine; Russian peasants were particularly ingenious manufacturers. Because Russia remained an agrarian country for longer than other European countries, the use of surrogate breads in times of food shortages continued well into the twentieth century. A study done at the University of Kazan in the 1890s found evidence of all these "breads": straw, birch and elm bark, buckwheat husks, pigweed, acorns, malt grains, bran, potatoes, potato leaves, lentils, lime leaves, cow parsley.[65] Also added sometimes were chaff, straw, and clay. Usually, they were combined with whatever ordinary flour was left, but as shortages got worse, the percentage of the additions crept up, often as high as 50 percent. Sometimes, things got so bad that these "foods" would actually be bought as well as gathered: pigweed might retail for twenty to seventy kopeks a pound.

There are accounts of the vile effects of these "breads" on the human body. Take pigweed bread, for example. Pigweed, or amaranth, is a common weed, growing often in vegetable patches. It is still some-times eaten as a green vegetable by wild-food enthusiasts, and its taste in this form has been described as "mild." As a bread, though, it was said to be unpleasantly insipid. According to the Russian peasants, pigweed bread gave them a terrible thirst: barely nourishing, it caused pain in the arms and legs, leaving them too weak to do their usual amount of field work. Yet still they ate it, since the alternative was death.[66]

Another revolting Russian "surrogate" bread was husk bread (*pushchnoi*), common in the Smolensk province in the 1880s. A contemporary observer, A. N. Engl'gardt, who lived in Smolensk, reported:

> Husk bread is made from unwinnowed rye, in other words, a mixture of rye and chaff is milled directly into flour and bread is made from this in the usual way. This bread is heavy and doughy and full of tiny needles of chaff. Its taste is not bad—about like ordinary bread. But its nutritional value is, of course, less. But the main drawback is that it is hard to swallow, and if you are not used to it you will simply find it impossible to swallow, or if you do swallow it, it leaves an unpleasant sensation in the throat, and makes you cough.[67]

If you were in any way frail, you would not be able to digest this turgid food.

Compared to this, a sturdy loaf of baker's bread made from real grain, whether short weight or full weight, must have seemed like manna from heaven. But food can provoke many different kinds of hysteria, and the thought of what could happen to bread during shortages was capable of provoking all sorts of unease, even when the shortages were mild and the bread still good. Sometimes the panic was based on objective concerns and needs. Fear would have been an entirely rational response, for example, to the problem of leaded wine; if anything, consumers did not panic enough about this.

Often, however, legitimate concerns about the food supply mutate into a kind of collective madness, creating an environment where the wildest accusation becomes suddenly plausible. The legitimate concerns vanish beneath the frenzied alarm. It has been said that true "food scares" are the preserve of the last twenty years, being the histrionic indulgence of "the healthy, wealthy, comfortable middle classes" whose world is so free of plague and war that they have to fabricate imaginary problems for themselves.[68] This is not true. It may have taken longer for these fears to travel in the time before newspapers, but they existed nevertheless, even in communities that were well acquainted with real hunger.[69] In Britain, there were periodic outcries that bakers' bread was not pure, which came to a head in the events of 1757.

The Great Bread Scandal of 1757–58

The year 1756 was a bad one for British wheat. Too much violent rain just before harvest was said to have levelled the wheat to the ground. The crop was smaller than usual and of much worse quality than the previous year. People complained that this soggy wheat wouldn't grind well, and the flour produced wouldn't bake well.[70] In response to the shortage, Parliament authorized a "Standard" bread to be made, stamped with a capital letter S. This was made with more bran than was customary, to extract more nourishment per penny, and it baked up darker than the standard loaves customers were used to. It sold cheaper too. To modern tastes, this bread sounds wholesome and good, if a little worthy. But it wasn't popular then. People associated bran-rich bread with poverty. They wanted bread that was whiter than white. As Hogarth wrote in 1753, "They eat no Bread of Wheat and Rye, but ... as white as any Curd."[71] The rage for white bread was silly enough at the best of times, but in times of bad harvest, as in 1756, it was unrealistic in the extreme. The only way bakers could make curd-white bread from poor flour was by adding alum to the flour. This particular year, some bakers seem to have overdone the alum, which provoked an unprecedented series of attacks on their trade.

Alum—the name given to a group of double sulphates that join aluminium sulphate with another sulphate (potassium or sodium or ammonium)—is an astringent, styptic and emetic with countless uses. From medieval times, it was a vital ingredient in the textile industry, as a "mordant," fixing dyes to fabrics. It has also been used externally as a deodorant and to staunch bleeding, especially from shaving. The old-fashioned gentleman's toiletries firm Geo. F. Trumper still sells a "block of alum" for closing nicks and cuts. Alum has also had many culinary uses. It has been used as a preservative, as a firming agent to create extra-crisp pickles and maraschino cherries, to harden gelatine, and as a flour improver and bleach. Bakers have used it in this last capacity since Renaissance times, though only to a significant degree since the eighteenth century.

The use of alum in bread makes sense only if we take into account the prestige of white bread. For a long time, the poor had aspired to eat the fine white manchet of the rich. White bread was seen as gentlemanly, whereas brown bread was yeoman's bread, marking one out as socially inferior. For this reason, those who had most reason to resent their social inferiority were the most demanding about their bread. Several observers in the seventeenth century describe the poorest classes travelling to market to seek out bread made from the finest white wheat flour, spurning rye bread as beneath them.[72] Almost no one wanted to be the sort of person who ate brown bread. There would always be the odd voice of wisdom in favour of wholemeal bread. A nonconformist vegetarian, Thomas Tryon, in 1683 spoke up for wholemeal bread as a natural food, good for digestion. He attacked the taste for white bread as "inimical to Health, and contrary to both Nature and Reason."[73] No one much listened. The taste for white bread continued, and with it the use of alum.

White bread was always more expensive than brown bread, because the bran was wasted and because the bread had to be made from wheat rather than the cheaper barley and rye. The only way to make white bread cheaper was to make it from inferior flour. But in this case, the bread would come out greyish and heavy rather than white, which ruined its potential as a status object. Hence, the use of alum, which could turn second-rate white flour into a light, white, porous loaf, at a low price. By the eighteenth century, the use of alum seems to have become more prevalent, apparently reaching a peak in 1756–57. Given the poor, sour grain, bakers used more alum than usual, which resulted in a "harsh" crumb and acrid taste. "The smell is raw and disagreeable and the taste has nothing of sweetness," said one unhappy customer.[74]

Faced with this bad bread, especially in the cities, disgruntled consumers came to see British bread as profoundly adulterated. So unpleasant was it, they thought, it must surely contain all kinds of shocking ingredients in addition to alum. Published in 1757, *Poison Detected or Frightful Truths* was the first, anonymous, baker-bashing tract (probably written by a Dr. Peter Markham, who went on to publish several other attacks on bread). The author claimed that "our bread, the universal basis of the food of all ranks and ages of people, is mixed with the most noxious and morbiferous matter."[75] He petitioned William Pitt, the prime minister, to prevent the "once venerated men of England" from being poisoned by bad bread.[76] The wickedness of bakers, in his view, caused more harm than the worst natural disasters: "Run over the gloomy roll of horrors; earthquakes, inundations, tempests, famine, lightning, fiery eruptions, venomous or savage animals, and deleterious plants; they will be found less baneful to human existence than . . . the secret craft of impetuous avidity."[77]

Alum, he argued, was a seriously hazardous substance whose frequent use "closes up the mouths of the small alimentary ducts and by its corrosive concretions, seals up the lacteals, indurates every mass it is mixed with upon the stomach, makes it hard of digestion, and consolidates the faeces in the intestines, so as to bind up the passages." In addition, alum caused heartburn. But what could you expect from a substance that the author of *Poison Detected* declared was actually "an extract from human excrement"? "Even the most stertorian stomach fastidiates the nastiness of a food made up with such a disgustful mixture." This was not all. The author also insisted that bread was poisoned with chalk and lime, which gave its crumb a "putrid alkalescence" on top of the "acrid acrimony" it received from the alum. The worst was yet to come. "There is another ingredient, which is more shocking to the heart and if possible more hurtful to the health of mankind"—"sacks of old ground bones," raked from charnel houses.[78] In other words, bakers were creeping around at the dead of night stealing dead men's bones. "Thus the charnel houses of the dead are raked to add filthiness to the food of the living." Another attack on bread of 1757—*The Nature of Bread Honestly and Dishonestly Made*—by a Dr. Manning also accused bakers of using bones, in the form of bone ash, to increase the weight of their flour, though he claimed that bones were stolen from dunghills rather than charnel houses.[79]

Needless to say, this last charge against the bakers was not allowed to pass uncontested. In 1758, a Bristol writer called Emmanuel Collins published a book called *Lying Detected*, dismissing the notion that bakers indulged in these cannibalistic practices. "At this rate," he wrote, "a man may happen to eat the bone of his own father's nose in a buttered muffin for breakfast."[80] These were fairy tales, and Collins cited the chant of the giant from *Jack and the Beanstalk*:

> Fe, Fa, Fum
> I smell the blood of an English Man,
> Be he alive or be he dead,
> I'll grind his bones to make my bread.

To Collins, the charnel house accusation was also culinarily implausible. If bakers really added bone flour to bread, "you might look into the oven and see it boil instead of bake; and your composition would come out broth instead of bread."[81] Modern historians have agreed with Collins that the charge of using human bones in bread was wholly fantastical. Yet thanks to the antibread tracts, "a shocked public firmly believed that their flour was mixed with dead men's bones and that the millers and bakers were a set of rascals."[82]

What of the other charges against the bakers? The author of *Poison Detected* had claimed that mealmen and millers added chalk and lime "in very considerable quantities" to augment the weight of the flour they grind, a charge echoed by Dr. Manning in *The Nature of Bread*.[83] Yet this was highly unlikely, as the chemist Henry Jackson wrote in 1758, in a level-headed work defending the bakers and millers (*An Essay on Bread*). Jackson evidently deplored the scaremongering of *Poison Detected*, rebuking the author as "he who alarms the populace with idle systems and conceits of poisons existing in bread." Jackson observed that if chalk and lime were used in order to increase weight, it would have to be in such huge quantities that "the bread would turn out so gritty that the very dogs would spurn it": the grittiness would be immediately discernible in the mouth. "The baker's profits would not be increased by such ridiculous substitutes," noted Jackson, plausibly enough.[84]

To make doubly sure, though, he conducted a series of experiments, trying to add chalk and lime to bread dough. He found that such dough was almost impossible to knead. Jackson's findings were confirmed by a twentieth-century historian of adulteration, Dr. Frederick Filby. Filby made three loaves of bread. One contained nothing but flour (10 oz), water (6 1/4 oz), yeast (1/2 oz), and salt (1/4 oz). "It was baked to perfection in 30 minutes and was later eaten with relish."[85] To the next loaf, Filby replaced 2 oz of flour with 2 oz of slaked lime. He found that the dough took twice as long to prove, was sticky and yellowish, and had a "loathsome" smell. It took twice as long to bake as the proper loaf and came out of the oven of short weight—an ounce lighter than the genuine loaf—and resembling "a flat crab" with "a shell of hard cement." A third loaf baked with 1 oz of chalk was also revolting, though not to the same degree. In other words, it just did not make sense that bakers should have used these ineffectual substitutes. Manning had claimed that the bad bread of 1756 must have contained such things as bone ash and lime because the crumb was so brown, crumbly, heavy, and brittle. Those who know about baking, however, have pointed out that these faults can occur without the presence of adulterating agents. Brownness could come from too much bran or poor yeast. Crumbliness and heaviness might be signs of too much or too little fermentation, or of using water that is too hot. A brittle crust may be a sign of soft flour.

The great bread scare of 1757, like so many food scares since, was one in which wild accusations were hurled around by interested parties, with scant regard for whether they were true or not, or where the stories might have originated. The belief that bakers used human bones in their bread continued for many decades, despite there being no evidence for it. Filby suggests that the insinuation may have arisen from bone ash being discovered on a miller's or baker's premises at various times. "It was undoubtedly used, not in flour at all, but to stop up cracks and holes in mill stones."[86]

81

This still leaves the question of alum. Even the bakers' defenders admitted that some of them did use alum, though they also pointed out that it was not quite so "disgustful" an ingredient as *Poison Detected* had asserted. It certainly never contained "human excrement." This mistake must have arisen because the manufacture of alum *did* sometimes involve human urine. In volcanic regions, alum can form naturally as alunite crystal. In the less than volcanic British landscape, though, alum was made from aluminium sulphate mined from shale—a type of sedimentary rock—at giant quarries. In order to get this aluminium sulphate to crystallize into alum, an alkali was needed. Stale urine is an alkaline solution and contains ammonium sulphate, and—crucially—it is universally available, so it fitted the bill. Local households would keep their urine especially to sell to the alum manufacturer's agents. Huge vats of urine were kept in alum-producing towns such as Whitby in Yorkshire, staling to an ammoniac stench. It is true that this may not be an appetizing thing to think about when considering bread, but it is not in itself a reason to reject out of hand the use of alum. Some people have consumed small amounts of urine without harm throughout history: rich Romans rubbed urine in their mouths to whiten their teeth; and "urine therapy," the drinking of your own, or someone else's urine, has long been considered normal in some parts of India and China.

The more pertinent question is whether alum, however produced, was dangerous to consume in the quantities in which it was used in bread. Dr. Manning had noted that in August and September 1756 there had been "a kind of universal distemper," a "habitual diarrhoea." As a doctor, he said that he had never seen "so many disorders among the robust and strong" as he had within the previous seven months.[87] Could this have been caused by the "concealed poison" of alum in bread? Even Henry Jackson, who denied that alum was quite as noxious as *Poison Detected* had said, admitted that it could be a purgative for some children and thought it "greatly to be wished" that alum in bread should be abolished altogether.[88]

Modern evidence suggests that swallowing alum can be hazardous. The International Labour Organization, which issues guidance on the toxicity of chemicals in the workplace, states that alum ingestion can cause "abdominal pain, burning sensation, nausea, vomiting," and its inhalation can result in "cough, shortness of breath, sore throat."[89] Taken straight, 30 grams of alum has been enough to kill an adult. In the case of bread, though, it was not taken straight, but in very dilute form. Dr. Peter Markham, the probable author of *Poison Detected*, calculated in 1758 that alum was added to bread in the proportions of eight ounces for every five bushels of flour, which was the standard weight of a sack of flour. There were 240 pounds to every sack, which, once water and salt and yeast were kneaded in, would have baked up to make around 350–360 pounds of bread. A standard daily ration of bread in the eighteenth century would have been one or two pounds, depending on what else was eaten with it. Based on these figures, an adult eating alum-adulterated bread might have consumed between 0.6 and 1.2 grams of alum per day—a fairly modest amount.

Whether it was still dangerous in these quantities is a moot point; the evidence of 1756 is that increasing the percentage of alum from this low level hugely increased its toxicity. As always with poison, some people must have been worse affected by alum than others, and children were likely to be worse affected than adults, poor children worst of all because bread made up such a large percentage of their calories. There is plenty of anecdotal evidence to say that people used to country bread—which was less likely to contain alum—suffered from wrenching dyspepsia when they ate whiter-than-white city bread. Accum noted that some doctors attributed many "diseases incidental to children" to eating adulterated bread while other doctors considered it "absolutely harmless."[90]

In 1758, the British government banned the use of alum in bread. If anything, though, the bakers' dependence on the substance seems to have escalated (as it did in the United States, too; an opposition to alum was partly what made Sylvester Graham urge his followers to bake their own wholemeal bread in the 1840s and 1850s). In 1851, Arthur Hassall examined various loaves purchased at random in different parts of London and found that they were all, without exception, doctored with alum, even one that was advertised as of "perfect purity, being warranted free from alum."[91] By 1857, a French friend could write to Eliza Acton that British bread was "noted both at home and abroad for its want of genuineness, and the faulty mode of its preparation."[92] How and why had British bread become so contaminated? There were undoubtedly economic causes; the periodic shortages of wheat and the constant pressure on "corn" prices in the late eighteenth and early nineteenth centuries put pressure on bakers to cut corners. Bakers, however, could only sell what they could get away with selling. The major factor contributing to the erosion of standards in bread making was the failure of British bread eaters—unlike their Parisian counter parts—to demand properly made bread. While they worried about fictitious bones in their bread, the real scandal was that they no longer knew good bread when they saw it.

Elizabeth David, the food writer, asked why British bread in 1977 was so poor and concluded that "scientists and their technological achievements have combined with commercial interests, compliant governments and the public's own indifference to give us the factory bread we now have. No doubt we deserve it. We certainly asked for it, and the milling-baking combines gave it to us."[93] Something of the same was true in the years following that disastrous harvest of 1756. Smollett in *Humphry Clinker* wrote that "The good people are not ignorant" of the adulteration of bread with alum, "but they prefer it to wholesome bread because it is whiter than the meal of corn. Thus they sacrifice their taste and their health . . . and the miller or the baker is obliged to poison them and their families, in order to live by his profession."[94] Many commentators observed that the buyer was "equally culpable" with the baker on the matter of alum.[95] If bakers did not meet the demand of their customers for white bread, they could go out of business. It did not help that the same assize that banned the use of alum in 1758 also made it much more profitable for bakers to bake white bread than brown. "In fact, the latter could only be made and sold at a loss and the bakers naturally did what they could to encourage the sale of white bread, even, according to one critic, by making the brown loaves so unpalatable that no one would buy them."[96]

Little by little, adulterated white bread became the norm for baker's bread, just as factory bread is the norm now. Some people, then as now, opted out of buying bread altogether, preferring to bake their own, either buying flour from a reputable dealer or grinding their own using a hand mill. The only person you could trust completely to make wholesome bread was yourself because there were no longer any enforceable compositional standards set by law. Over the years, the assize of bread had been altered many times, and it would eventually be abolished in 1822. The author of *Poison Detected* criticized the government for not making "severer laws" on bread. If only bread could be baked in one communal oven, as it supposedly was in the city of Genoa![97] In 1819, another anonymous critic made a similar point. "Much and proper precaution is used to secure to the publick the just weight of the loaf; but why should not competent persons be equally authorized to analyse its composition? The expense would be insignificant, the benefit of the highest importance to the community."[98]

The truth was there *had* once been such people analysing the quality of foodstuffs in Britain, but they belonged to a different preindustrial era—the feudal world of trade guilds. The example of bread shows that food laws by themselves are not enough, unless backed up by experts who know how to enforce them. Just as the rise of modern wine was partly caused by new levels of expertise in the matter of wine quality, the decline of British bread was linked to a terrible decline in Cyrus Redding's "perfect acquaintance with that which is good." The sad thing was that, unlike with wine, such expertise *had* once existed in Britain. It had been passed down by the craft organizations that governed trade in medieval Europe.

Guilds and the Guarantee of Good Food

In *Capital*, Karl Marx argued that:

> The adulteration of bread and the formation of a class of bakers that sells the bread below the full price, date from the beginning of the eighteenth century, from the time when the corporate character of the trade was lost, and the capitalist in the form of the miller or flour-factor, rises behind the nominal master baker.

Marx was wrong to suppose that adulteration was new. "For they poison the people secretly and often," wrote William Langland in

Piers Plowman, as long ago as the fourteenth century.[99] Human greed is a constant. But Marx was right that competition between bakers to undercut each other exacerbated adulteration, whereas the old guild system had done much to prevent swindling. From 1307 to 1509, there was a company of white bakers and a company of brown bakers in London. Each saw it as their corporate mission to ensure that the quality of their respective breads remained high. If an individual baker should bake a loaf of bread that was of poor quality or tampered with in some way, it reflected badly not just on that individual but on the entire company. Joining a guild was usually an expensive and burdensome business, offering a status that you would not want to throw away for the sake of cheating some customer out of a few farthings. This created serious disincentives to swindling.

Guilds governed much of the trade of the cities of Europe from the eleventh century onwards. The various specialized guilds were, by definition, each jealous of the reputation of their particular trade. It has been said that "the guild prided itself on letting nothing leave its shops but finished products, perfect of their kind ... Not only fraud, but the very suspicion of fraud was rigorously excluded."[100] Guilds laid down very stringent rules on goods for sale, to preserve the honour of the craft. For example, all goods were to be sold under exact names. The oil of Puglia should not be mixed with the oil of the Marche, because this might make people think that all oil merchants were slapdash.[101] Fishmongers should not use seaweed to "freshen" old fish, lest customers draw the conclusion that such deception was common to all fishmongers. It was "strictly forbidden, under penalty of a fine or expulsion, to sell damaged meat, bad fish, rotten eggs, or pigs which had been fed by a barber-surgeon who might have fattened them on the blood of sick people."[102]

In some ways, guilds had a stifling effect. They were intensely hierarchical and gave rise to ever greater specialization and endless squabbling over exactly who had a right to which area of trade. The bakers squabbled with the confectioners, "the cooks with the mustard makers ... the dealers in geese with the poulterers etc."[103] Those who think that progress is the most desirable thing in food might argue that one of the effects of the guilds was to dampen creativity.

In the eighteenth century it took a long time for the restaurant to become established in France, partly because of interminable squabbles between the guild of hot broth sellers and the guild of cooked sheeps' trotters. Because guilds usually held a monopoly on their trade in that particular town, there was little incentive for improving and innovating.

On the other hand, guilds were, on the whole, excellent at maintaining quality and tradition in food. In the town of Maine in France, a butcher was not allowed to display a piece of beef on his stall unless two witnesses could testify that they had seen the animal brought in alive.[104] At Poitiers, butchers had to undergo a physical and moral examination to ensure that they were not scrofulous, or suffering from scurvy or bad breath, nor morally unsound. The guilds of British cities developed similar rules. Anyone who broke these rules could be expelled from the guild. This is not to say that guild members never sold bad food. Court records from London in the seventeenth century show that members of the Worshipful Company of Fishmongers were occasionally indicted for selling "stinking sturgeon" and "unsweet cod."[105] But at least the guild rules meant that there were robust standards for food, even if these were not always met.

Not everyone producing food and drink belonged to a guild. Many people acted as their own butchers and bakers, killing their own pigs at home and making them into sausages, bacon, and pork. There were those too, who sold nonessential foodstuffs—such as the "cheesers" and "fruiters"—whose trade seems to have been less organized or else less well documented than that of the butchers and the bakers. In addition, there were those who drifted informally in and out of food production, indistinct groups of sauce-makers or cooks who escaped the control of the guild; glaziers' wives who had a sideline in brewing or cordwainers who made a few extra shillings from selling fish.[106] Sometimes they were clamped down on. A York ordinance of 1424 specifies that "the wives of any other artisan shall not bake, boil or roast fowl in public shops, for sale, unless they are competent to do so."[107] In other words, artisans' wives must not try to moonlight by selling food unless they knew how to cook. Doubtless, the guilds had a selfish interest in keeping such unlicensed provisioners out of the trade; but they also had an interest in making their own food as wholesome as possible.

In the Middle Ages, both the guilds and the law treated food production as not merely a profession but a duty. If you were in one of the victualling trades, part of your job was guaranteeing that everyone had access to high-quality food, and enough of it. Hence, there were local laws not just to prevent bad food from being sold, but to ensure that good food was sold and available to those who needed it. Bakers were sometimes penalized for failing to bake, as happened in York in 1485. In some cities, the victualling guilds were taken over by the government, to make sure that the poor wouldn't starve in hard times. The commune of Florence had a monopoly on salt; Rome had a monopoly on the city's fisheries. There were various restrictions on professional victuallers to prevent them from stockpiling or "engrossing" food for themselves—for example, a York ordinance of 1497 forbade bakers from buying corn in the market before midday, to make sure they did not monopolize the grain supply.

These laws were made in the interests of the consumer, but generally the interests of the guild coincided with those of customers. The consumer wanted guarantees of quality, and the guild wanted the high reputation that came with providing these guarantees. A guild structure more or less guaranteed sales to those who had achieved membership. Modern food manufacturers are engaged in a constant struggle to undercut their rivals and to reposition their brand with new products. Guilds were very different. Their job, rather, was discovering the secrets of making their particular product as well as they could—whether it was meat pies or gold rings—and then jealously guarding these secrets against the world.

In today's commercial world, attempts by government to improve the food supply can be interpreted negatively by the many food producers, as an infringement on their freedom to trade as they see fit. In the feudal world of guilds, however, government and trade mostly worked to a common purpose. Indeed, subgroups of guild members took on an official character, assuming the role of policing food and drink in order to guard against malpractice. Some of the decline in British food quality, including that of bread, is tied to the fact that the guild system eroded so early; it was already waning in the age of Shakespeare. In France, by contrast, the self-policing role of guilds continued right up to the revolution of 1789. Although the French despair periodically about their bread, something of the guild mentality survives in France to this day—the communal pride in a great loaf, the notion that, as the abbé Galiani once said, "bread belongs to the police [meaning government] and not to commerce."[108] British food has belonged to commerce for centuries now. The self-policing guilds gave way to the unregulated trade of the grocer's shop, and from there it was a short leap to the twentieth-century supermarket. The irony was that, once upon a time, the "grocers" were the people whose job it was to prevent swindling.

The Food Police: Conners, Pepperers, and Grocers

In France, there was a peculiar subgroup of officials known as the *langueyeurs de porc*. Their job—a specialized one even by the standards of the day—was to examine pigs' tongues to see if they showed any signs of leprosy, to protect the consumer from diseased meat.[109] Other testers had a much broader remit. For example, the quality of British ale in the late medieval period was superintended by a class of men known as "conners." The first ale conner was appointed in London in 1377. His duties were to taste the ale of "any brewer or brewster" and pronounce on its quality. If it tasted "less good than it used to be," the ale conner was entitled to lower its price. Brewers must have often felt tempted to bribe the ale conner to give a better rating of the ale than was justified, though such corruption was expressly forbidden. Ale conners were banned from changing their opinion "for gift, promise, knowledge, hate, or other cause whatsoever." If the ale was mouldy, or too thin, or somehow unwholesome, the ale conner was expected to say so.[110] Equally, if an ale was fine and well made, he must speak up, even if he personally disliked the brewer who made it. We no longer have ale conners. Now this role is filled, insofar as it is filled at all, by health and safety inspectors, on the one hand, and food and wine critics, on the other. The wine critic Robert Parker may lay claim to being the true inheritor of the ale conners, a man who can, by his incorruptible palate alone, ruin or enhance the business of a vineyard.

Then there were the antifraud brigades: guilds of measurers and verifiers throughout Europe, charged with the responsibility of preventing frauds of weights and measures. In France, these operated as the guild of *mesureurs*, who verified the capacity of barrels, jars, and so on, checking that they were as big as they said they were. In Britain, their place was taken by mysterious bodies known variously as pepperers and spicers, garblers and grocers.

The first record of the guild of pepperers—the *Gilda Piperarorium* or "mistery of pepperers"—dates from 1180.[111] These men were not just pepper merchants (they dealt in all manner of dried goods or "spicery," such as sugar, dried fruits, and alum, in addition to pepper). Pepperers were a group singled out by the king for special honours and obligations. Out of all the food guilds, it was the pepperers who were granted the custody of the king's weights, and later the office of sifting through commodities to check for adulteration. Why the pepperers? Because of all the traded foods, pepper was the most prestigious, and it needed to be weighed in large amounts. It was not just something fiery to grind over food. Pepper was used to pay taxes, rents, and dowries. The price of pepper was a barometer for the state of business in general. "Pepper was always more involved in trade than any other spice."[112] And unlike salt, which was relatively pure, pepper had always been prey to swindling. Pliny writes of pepper adulterated with Alexandrian mustard and juniper berries as early as the first century A.D.

Pepperers and spicers appear in the record books as the king's weighers as far back as 1285, in the reign of Edward I, when it was decreed that "The king shall have his weights in a certain place, or in two, three or four places within the City" and that all merchandise weighing more than twenty-five pounds shall be weighed by the king's weights by weighers who had been sworn in by the king. The king had different weights for weighing in bulk and weighing in smaller amounts. "Spicers" based in Cheapside dealing in small amounts of cinnamon, ginger, and so on—*les sotils choses* or "small things"—were entrusted with the Small Beam of the City (*peso sotil*), which weighed things in light twelve-ounce pounds. The Great Beam of the City—the Gros Beam—was given to the pepperers of Soper's Lane, who used heavier weights than the spicers—the *peso grosso* or *avoirdupois* weights that correspond to our modern pound. Because of this, the community of pepperers came to be known as the *Grossarii*, the custodians of the Gros Beam.[113] In 1345, they joined forces with the spicers to form the Fraternity of St Anthony, named after a church on the junction of Soper's Lane.[114] In 1373, this powerful brotherhood renamed itself the mestres de la Compagnie de Gr'ssers—the Grocers' Company.

It wasn't long before the Grocers' Company was granted the office of *garbling* or cleansing all spices before they were sold to the public. These "garblers" have been described as "the first guardians of the public food and health."[115] The word "garbler" comes from the Arab *gharbala*, to sift or select, and this is exactly what the garblers did, using a series of gradated sieves to filter "pepper, ginger, cinnamon, etc." to detect impurities such as gravel, leaves, twigs, and other dross, and using their senses and expertise to check that the spice was genuine.[116] It was illegal to mix old spice with new, or to moisten saffron, cloves, or ginger to increase the weight. Every drug and spice that landed at the London docks had to be checked by a garbler and verified before it could be sold. A document of 1380 states that it was strictly forbidden to sell "merchandise of grocery" in an "uncleansed state" until it had been "garbled by a man appointed for the purpose by the said grocers."[117]

Like the ale conners, garblers had to take an oath to their integrity:

> You shall swear that you shall well and honestly behave yourself in the office of garbelling, within the City of London, without stealing, embeazelling, or unlawfully or unhonestly conveying away any part of such spices as are left to your charges in any merchants house or elsewhere. You shall, as much as in you shall lie, garbell and cleanse all manner of spices, drugs and merchandise, justlie, trulie and indifferently according to your skilled judgement without respect of any person or persons whatsoever.[118]

Garblers had a self-image as sentinels against swindles. In the early seventeenth century, their office came under attack from the East India Company, the main importer of spices, which sought to exempt itself from having its goods garbled. The garblers, however, protested that without them "the consumers of garbleable commodities . . . and all other consumers of spices and drugs, are those who will be unavoidably injured, and ruined both in their persons and estates by abuses and frauds."[119] They had been preventing "frauds and abuses," they alleged, for three hundred years; it was wrong for the East India Company to seek indemnity from being garbled. This clash was a taste of things to come. Big corporations such as the East India Company would eventually sap the power of the guilds from which the garblers came. And the garblers were right. It does seem that spices were far more systematically adulterated by the late eighteenth century than they had been when the garblers were in charge.

Garblers were not just acting from the goodness of their hearts, though. They were well remunerated for their services. At seventeenth-century prices, they charged two shillings a bag for garbling pepper, three shillings and six pence per hundred pounds of nutmegs, two pence a pound for cloves, eight pence per hundredweight for bay berries, cumin, coriander, caraway, almonds, and rice, and twelve pence per hundredweight for ginger and pimento.[120] The fact that garbling was compulsory, and a monopoly, could give rise to corruption. King James I (1566–1625) received numerous complaints against the garblers. Petitioners complained that in many cases, garblers would set their seal on casks without properly checking through the goods. Spices coming in from Holland were known to have come ready garbled, yet the garblers would still charge for their services on Dutch spices.

In 1613–14, a number of petitions to the king argued that the garblers were no longer up to the job. Traditionally, garblers had been responsible for policing drugs as well as spice—there was no clear distinction between the two. But now they seemed ill-equipped to judge the purity of drugs. Corruption in the sale of drugs was becoming rife, and detecting the subtle adulteration of essential oils—with oil of turpentine, say, or expressed oil or spirits of wine—required the skills of a chemist, not a garbler. In 1617, the Apothecaries' Charter was granted and, little by little, the influence of the grocer-garblers as policers of swindling waned. The control of drugs passed to the apothecaries or pharmacists, and the control of food and drink was left with no special body to oversee it.

The state would continue to intervene to prevent food and drink fraud, but its main interest was in safeguarding the Exchequer rather than the stomachs of its citizens. With the wane of the garblers, there was no one to protect the interests of the consumer of food against the swindlers until the issue of adulteration came before Parliament in the 1850s, when the state was finally persuaded to start taking the risk to public health from adulterated food seriously. By contrast, over the same period, the reputation of grocers moved from being that of guardians against food fraud to being the worst perpetrators of it. In the nineteenth century, no group of sellers was so mistrusted. The story of the battle against food adulteration was to become the story of the battle against the duplicitous grocer.

3

GOVERNMENT MUSTARD

*We ask for bread, and we receive a stone; for coffee, and we receive
chicory; for chicory, and we receive burnt carrots and powder of
dried horses' liver; for oil of almonds, and we receive prussic acid,
to heighten the enjoyment of the dessert by adding a little risk to it.*
                                        —*The Times*, 3 March 1856

One of the besetting questions about adulteration is: why do people tolerate it? In 1868, George Eliot revived one of her characters, the idealistic Felix Holt (from her novel of that name), to pose this question. It was the year after the Second Reform Act, which had given one and a half million working men the vote. There had been heated debate among the Whitehall establishment as to whether the working classes were intelligent enough, or good enough, to have the franchise. To Felix Holt, though he wanted democratic reform, it was clear that the majority was "neither very wise nor very virtuous." If they had been so, they would not have put up with the swindling they suffered:

> Any nation that had within it a majority of men—and we are the
> majority—possessed of much wisdom and virtue, would not tolerate
> the bad practices, the commercial lying and swindling, the poison-
> ous adulteration of goods, the retail cheating, and the political bribery,
> which are carried on boldly in the midst of us. A majority has the power
> of creating public opinion. We could groan and hiss before we had the

franchise: if we had groaned and hissed in the right place, if we had discerned better between good and evil, if the multitude of us artisans, and factory hands, and miners, and labourers of all sorts, had been skilful, faithful, well-judging, industrious, sober—and I don't see how there can be wisdom and virtue anywhere without these qualities—we should have made an audience that would have shamed the other classes out of their share in the national vices.[1]

Holt is telling his audience of working men to make full use of the opportunity they have been given. But his (or George Eliot's) words show just how difficult it was for the bulk of the population to do anything about adulteration—even to groan and hiss—when it did not have the franchise. Before they had the vote, working men and women (the latter did not get the vote until 1928) had to swallow many things they did not like, and not just bad food.

Despite an increasing awareness that adulteration was as prevalent as Accum had alleged, there was no popular crusade against it in the decades after 1820. The commercial defence of adulteration as something almost respectable, which Accum had so deplored, continued as if he had never written. French social critics at this time often accused the English—perfidious inhabitants of *perfidious Albion*—of hypocrisy. Judging from the attitude of the British government concerning the safety of what people ate, the French had a point. As far as politics went, the working classes were not deemed responsible enough to vote, but kept in a state of childlike dependence. Yet in matters of commerce, they were treated as responsible enough to make sophisticated judgements about the safety of the food they bought. It was a "buyer beware" culture, which foisted huge responsibility onto a populace that lacked even basic democratic rights.

The food culture we now live in is the opposite of this. The rule now is, for the most part, not buyer beware but seller beware. Around the turn of the new millennium, a strange new trend emerged in restaurants in both America and Britain. If you ordered a steak or a burger and asked for it to be done rare or medium rare, you might find yourself with a piece of paper thrust on your place setting. Unless you were prepared to sign a legal disclaimer accepting that the restaurant could take "no responsibility" for the safety of meat cooked in this manner, you couldn't have your burger pink in the middle. These burger disclaimers, which annoyed numerous impatient diners, were an attempt to redress an imbalance. They were getting at the question of who is responsible for the food that consumers put in their mouths. Restaurateurs—who often feel burdened with responsibility from all sides, from environmental health officers, from fussy customers, from government red tape—were trying to shift a little of it back on to the individual. In our seller-beware food culture, the government, the press, advertising standards agencies, and consumers all exert pressure on food-sellers to stick to their promises and to be clear about what those promises are.

It was very different in mid-Victorian Britain. The collapse of the guild system had left a vacuum. Who would take responsibility for preventing swindling? Not the government: except for intervening to seize badly tainted meat, and for policing heavily taxed items such as tea and coffee, successive Victorian governments were reluctant to intervene in the case of food. Nor the press: though it publicized bad cases of adulteration *after* they had happened, it did little to provide the consumer with information that could protect against buying falsified food. Nor the sellers themselves: one contemporary observed that when accused of adulterating food, the shopkeeper "endeavours to shelter himself, and to excuse his dishonest practices, under the assertion that the public 'likes it' and '*will have it.*'"[2] So who was responsible for preventing swindling? The individual consumer, it seems: the person least equipped to do anything about it. In the "buyer beware" culture, the buyer had a great deal to beware of.

Demon Grocers

In the decades after Accum, sellers of food in England acquired an increasingly bad name, and none more so than grocers. The days when the grocers—through their association with garbling—could pride themselves as guardians of public health were long gone. Now their reputation was that of sinister crooks, conspiring in private to defraud their clientele. In 1851, *Punch* magazine wrote of "The

Grocer-Imp, who enriches his chocolate with brick dust; and government mustard with a morning draught conveys the materials of a vault."[3] *Punch* also carried cartoons attacking grocers. "The Great Lozenge-Maker" was one, a picture of a scary skeleton mixing up arsenic and plaster of Paris and turning the deadly mixture into "bon-bons for juvenile parties." When G. K. Chesterton published his "Song Against Grocers" in 1914, he was crystallizing a view of the profession that had been commonplace for most of the nineteenth century.

A British grocer's shop, 1866.

God made the wicked Grocer
For a mystery and a sign,
That men might shun the awful shops
And go to inns to dine . . .
He sells us sands of Araby
As sugar for cash down;
He sweeps his shop and sells the dust

The purest salt in town,
He crams with cans of poisoned meat
Poor subjects of the King,
And when they die by thousands
Why he laughs like anything.

"JOHN, have you sanded the sugar and watered the milk and molasses?"
"Yes, sir."
"Then you may come in to prayers."

From *Life*.

A cartoon depicting the common belief that grocers routinely added sand
to sugar. In fact, Hassall showed that this was one swindle that grocers
were not guilty of.

Yet by the time this poem appeared, the view it presented of the
demon grocer was already out of date. In the early years of the twentieth century, there was clear evidence that sugar was not bulked

out with sand, except in the popular imagination. Even in the worst grocer's shop of the late Victorian age, salt had invariably been discovered to be just salt and not sweepings. In fact, by the time Chesterton was writing, English food was much less adulterated than it had been for a large part of the nineteenth century. This chapter will explain how the change occurred, and how much it owed to the tireless efforts of one strange and unhappy man.

The true heyday of the demon grocers was from the 1810s to the 1850s, when all the staples of the English grocer shop were routinely falsified and padded and sold under weight, and although their customers suspected as much (and although Accum had confirmed many of their suspicions), there was little or nothing they could do about it. As well as the pepper traded by their medieval counterparts, the main products sold by a grocer or general shop were flours, sugar, and spices; cheese, butter, bacon, and salt fish; and, above all, tea and coffee. The precious powders on the grocer's scales were often mired in double or even triple adulterations. Ground coffee was almost universally thinned with chicory powder, the chicory itself cheapened with one or several of the following: roasted wheat and rye flours, burnt beans, acorns, mangelwurzel, sawdust, or a powder made from roasted dried carrots and parsnips. To disguise the adulteration, it needed to be darkened with burnt sugar. When making coffee, "never be tempted to try the deadly mixture sold at the grocers," advised one Victorian cookery writer.[4] Ginger was bulked with ground rice and sago, then, to make up for the lost pungency, doctored with cayenne; cayenne itself was often padded with further ground rice, or mustard husks, or dead sawdust, before being brightened in colour with turmeric if you were lucky or red lead if you were unlucky. No wonder the Victorians were so suspicious of spicy food.

Then there were all the sneaky ways in which unscrupulous grocers contrived to sell their customers short. One grocer's apprentice who had learned his job in Berkshire remembered how his master, a strict Methodist but no less a swindler for that, taught him the tricks of the trade.[5] The key principle was to con the customer out of as much as you could get away with. "You must make the ounces pay" was the master's mantra, especially when it came to items such as bacon, which the grocer needed to sell cheap to keep his customers, in the same way that modern supermarkets sell "known value items" such as milk and bread at rock-bottom prices to lure people into the store. The idea was to estimate how much swindling the customer would put up with and then add an extra ounce or two to every purchase. If a customer bought a hock of bacon weighing 6lb 7oz, at 4 1/2 pence per lb, they would be charged for 6lb 9oz at 5 pence a pound. Result: 2 1/2 pence extra profit for the grocer. When sugar was weighed, this apprentice recalled, "some was always spilt loose on the scale opposite the weight, which remains in the scale, so that every pound or so is a quarter of an ounce short." It was the same with cheese and butter. "On the cheesemongery side we were always blamed if we didn't keep the scale well wetted, so as to make it heavier on one side than the other—I mean the side of the scale where the butter was put—that was filled, or partly filled with water, under pretence of preventing the butter sticking, and so the customer was wronged half an ounce in every purchase."[6]

Needless to say, not every grocer was dishonest. There were honourable grocers too, who sold fair weight and did their best not to add to the impurity of what they sold. It must have been very distressing for them that the goings-on of their swindling counterparts gave the whole profession a bad name. After he left his apprenticeship in Berkshire, the grocer mentioned above got a job at another grocer's shop, this time in Yorkshire, an honest one, and "had to learn my business over again, so as to carry it on fairly."[7] In London, moreover, if you had money, you could go to one of the society grocers, either Fortnum's in Piccadilly or Crosse & Blackwell in Soho Square.[8] For the rich, this was the first golden era of the English shop, when the gaslight that Accum had fought to legalize shone on fantastical window displays of luxury goods in exquisite containers. As well as every sort of basic grocery, Fortnum & Mason sold picnic hampers of cold duck and lobster salad, truffled birds and champagne. Queen Victoria bought concentrated beef tea from Fortnum's, and it also supplied many of the gentlemen's clubs of the West End: the Garrick, the Athenaeum, Brooks's, the Carlton. Meanwhile, from Crosse & Blackwell you could buy pâtés and pasta, crystallized fruits and chocolates, jams, syrups and essences, oils and vinegars, twenty-five different pickles, and forty different sauces. As Friedrich Engels (one German immigrant to England who did not become a Frederick) wrote in 1844, "In the great towns of England everything may be had of the best, but it costs money."[9] Not everything sold at these grand shops was strictly unadulterated—of which more later—but the overall quality was infinitely better than at the average London grocer. The sad truth was that food swindling was something that disproportionately affected the poor.

Poverty and Adulteration

If you are a reasonably affluent First World citizen and you care about what you eat, you can create for yourself a fairly rich and pure diet. You can buy swags of fresh green Swiss chard from a farmer's market, straight from the farmer who grew it; you can steam it and drizzle it with unfiltered olive oil and scatter it with flakes of the cleanest sea salt; you can source chickens that have led a carefree organic existence; you can pay £4 for a loaf of the best nonbleached natural levain bread; you can drink pesticidefree, locally produced apple juice. It is different if you are poor and living in the same First World city. Unless you are unusually lucky or unusually persistent, your food is liable to be more corrupted than the food of the rich. Your meat is more likely to come pumped with hormones and water. Your bread is more likely to be bleached and enzymed and generally depleted. Your fat is more likely to be hydrogenated. Juice is too expensive, so your children drink squash laced with colourings and sweetener. This situation is unfair, though there are ways out for the lucky or the very determined. Perhaps you live near a good food co-op, or grow your own vegetables on an allotment, or take the time to buy big bags of healthy lentils and rice from an Asian grocer. But the disparity is still there. The rich can eat unadulterated food without much bother, whereas for most of the poor, it is a constant effort.

It was the same, only worse, in the 1840s, when the poor were so much poorer and faced the intolerable bind of being swindled for cheap food that they couldn't really afford even at the low prices at which it was sold. In the late 1840s, the journalist Henry Mayhew went onto the streets of London to observe at first hand the working lives of the London poor. He wrote up his observations in a series of reports for the *Morning Chronicle*, later published as *London Labour and the London Poor* (1851). Some of his most shocking reports concerned the dreadful state of the food of the poor. "When we think of the short weights and measures," Mayhew wrote, "and the quality of the articles supplied, we shall readily perceive how cruelly the poor are defrauded, and that if they are underpaid for what they do, they are at the same time fearfully overcharged for all they buy."[10]

If the middle classes bought food from markets and bakers and butchers and grocers, the labouring poor were often forced to patronize street hucksters, hawkers, and costermongers, whose prices were lower but whose swindling was even greater. Where grocers stole an ounce here and an ounce there, hucksters were far more blatant in their thieving. Mayhew discovered that the pound weight used by hucksters was usually four ounces short, and sometimes even as much as eight or ten ounces, meaning that for the poor, a "pound" of food could be in reality less than half that, and what seemed cheap was merely dishonest. The street pint was at least a third short. Mayhew noted that "as a body, the costermongers," who took their name from a type of apple but who actually sold an eclectic range of provisions, "rank high amongst the criminals of the country."[11] Based on prison records, there was one criminal in every 247 for butchers, but one in 86 for costermongers and hucksters.

Reading Mayhew, you see that swindling was a normal part of their trade, not something limited to a few crooks. Not everything the "costers" sold was bad; Mayhew was impressed with the "excellence" and variety of home-grown fruits and vegetables during the summer months at "the green markets of the metropolis," the bunches of watercress, the pink radishes, the asparagus and broccoli and plums; the gooseberries and strawberries, raspberries and currants. These delicious things were then sold on by costermongers. But even good food was often sold fraudulently by the time it reached the poor. Costermongers sold plums in a quart container with a false bottom. Cherries might be sold at the knock-down price of a penny a pound, the catch being that "a pound" really meant five ounces. Poor customers had to put up with this swindling because, first, they probably did not own any weights of their own, so they had no way of testing for sure that they had been "had," and second, because there was nowhere else to go. No one sold things cheaper than the costermongers.

Their working hours made the poor even more prey to swindling. As Engels angrily wrote in *The Condition of the Working Class in England*, it was common practice for a workman to receive his weekly wages only on a Saturday afternoon. As a result, he might not manage to get to market to buy food until five or seven o'clock in the evening. Thus, what Engels called the "property-holding class" got the "first choice" of food in the morning "when the market teems with the best of everything." "But when the workers reach it, the best has vanished, and, if it was still there, they would probably not be able to buy it. The potatoes which the workers buy are usually poor, the vegetables wilted, the cheese old and of poor quality, the bacon rancid, the meat tough, taken from old, often diseased cattle, or such as have died a natural death, and not fresh even then, half decayed."[12]

The business of Saturday night shopping increased the likelihood of buying inedible food. In the dark, it was hard to gauge the quality of what was on sale, and hawkers exploited this. Fish-sellers would save "rough" fish to sell on a Saturday night, using candlelight to make a darkened, smelly old mackerel look fresh and new.[13] In the north of England, some street-sellers even used a little red paint to touch up the fish's gills, since red gills were a sign of freshness.[14] This trick worked better in fading light than in the full glow of day, when the fishy makeup must have been all too obvious. For diseased meat and cheese, a devious technique known as "polishing" went on, whereby a putrid surface was covered with something fresh. Old meat was "polished" with a layer of fresh fat, and the cut surface of old cheese with a layer of fresh cheese. Similarly, old salted butter was covered with a layer of fresh sweet butter. Some Saturday night tricks were even more ingenious. In Manchester, factory workers might sometimes buy a coconut for a great weekend treat. Most customers would not spend money on such an extravagance without first shaking the coconut, to check that it was full of milk and therefore fresh. But devious sellers got around this by taking old, rancid, milkless coconuts, piercing them, filling them with water, and sealing them with a blackened cork, to match the brown of the coconut shell. Another wicked dodge was boiling oranges to make them weighty and shiny. By the time people got them home and found that the segments fell apart in their hands in disappointing cooked lumps, it was too late.

All of these were classic adulterating tricks, with sellers pulling the wool over the eyes of innocent buyers. But there were also cases where the Saturday night buyers must have known in their heart of hearts that what they were buying was likely to be no good. Engels observed that many workers chose to buy their family's food as late as ten o'clock or midnight because they knew they would get the cheapest food of all; but they must surely have realized that the price of this cheap food was its rottenness.

> As nothing can be sold on Sunday, and all shops must be closed at twelve o'clock on Saturday night, such things as would not keep until Monday are sold at any price between ten o'clock and midnight. But nine-tenths of what is sold at ten o'clock is past using by Sunday morning, yet these are precisely the provisions which make up the Sunday dinner of the poorest class. The meat which the workers buy is very often past using; but having bought it, they must eat it.[15]

Periodically, meat inspectors would seize tainted meat being openly offered for sale. Engels cites a case of sixty-four stuffed Christmas geese "which had proved unsaleable at Liverpool and had been forwarded to Manchester, where they were brought to market foul and rotten."[16] In fact, the selling of diseased meat was one of the few areas of British food that was policed by the law. Mayhew describes meat inspectors patrolling the markets to check for diseased and tainted meat. If a diseased animal should be found, it was condemned, handed over to the police, and its flesh boiled down under police supervision, to eliminate any chance of someone eating it. In August 1844, twenty-six tainted hams were seized at a dealer's in Bolton and publicly burnt. The dealer was fined. In January 1844, eleven Manchester meat-sellers were fined for having sold tainted meat. Yet despite the law's intervention, the selling of bad and putrid meat carried on.

This was partly because the selling of cheap bad food operated as a kind of black market, with buyer and seller colluding together against the law. At a modern street market, when a seller offers you a "designer" perfume for an implausibly low price, you know—if you ponder the matter for an instant—that it must be either stolen or fake. This is a buyer-beware scenario. If you go ahead and buy it anyway, you are complicit with the swindle. Perhaps you tell yourself that the real swindle is the high price of perfume in the official shops. A similar ethos must have operated in the food markets of Victorian England. When an expensive meat such as Christmas goose was sold impossibly cheap, this was a signal that something was wrong. If buyers went ahead and bought it anyway, they ought to have known that the seller was probably not honest. They would also have known that the price of regular, respectable middle-class goose was simply too high. Unlike the illicit perfume, however, which might only make you smell bad, the illegal meat could make you extremely ill.

In the poverty-stricken city of the 1840s, food buyers and sellers had a curious relationship—close but based on mistrust. Both buyers and sellers seemed to be victims of a market that was beyond their control. Those selling to the poor were, for the most part, themselves just as poor. The existence of a working-class coffee-seller could be precarious. Having shelled out money to garnish his stall with boiled eggs, watercresses, bread and butter, and fruit cake, the coffee-seller would water down the brew he sold as far as he possibly could with both chicory and water—it was typical to use ten ounces of coffee to make five gallons of the drink (the equivalent of 22.75 litres), which is about four times as dilute as most modern recommendations—but still struggle to make a living from the poor workmen who bought the coffee at around 1 1/2 pence per cup. To make the drink still cheaper, it was said that in the East End a group of "liver bakers" set up shop, selling baked powdered liver to make the coffee go further:

These men take the livers of oxen and horses, bake them, and grind them into a powder, which they sell to the low-priced coffee shop

105

keepers, at from fourpence to sixpence per pound, horses'-liver coffee bearing the highest price. It may be known by allowing the coffee to stand until cold, when a thick pellicle or skin will be found on top.[17]

If true, this disgusting liver coffee was simply the final economy for the working-class coffee-seller. It was a constant toil to make a profit, given the efforts of rivals and the poverty of the customers. Some sellers saw their customers as the enemy, terrible "screws" who had barely any money to spend and were always trying to beat down prices. One seller told Mayhew that hucksters were often forced to sell food at cost price. "The people haven't got money to lay out with them—they tell us so; and if they are poor, we must be poor too."[18]

From the seller's viewpoint, in this impoverished economy, a little swindling could seem like personal justice—a way of clawing back a living for themselves and exacting some revenge on their customers. Victorian fishmongers used different weights depending on the personalities of their clientele. If they considered a particular purchaser to be a "scaly cove" (in other words, a tight-fisted buyer), they would be more likely to use short weights and "always take care to have the laugh on their side."[19] More generous or less troublesome customers— "jonnocks"—might be rewarded with accurate weights. Often, though, the undercutting was employed on all customers, without favouritism, because the hucksters knew that, perversely, it was the only way to keep their custom. It was only through undercutting that they could afford to sell cheap; it was only through selling cheap that they could compete with their rivals. One huckster confessed to Mayhew that "We're all trying . . . to cut one another down, because we all want a livelihood, and unless we did cut one another down we couldn't get it."[20]

However much the sellers shifted the blame onto economic circumstances, however, this undercutting was not a victimless crime. It had a direct effect on the health of the poor. Engels noticed that the "adulterated and indigestible" food of the poor gave them seriously "impaired digestion."[21] Especially cruelly, undercutting foisted adulterated food on the poorest of the poor in the workhouses, prisons, hospitals, and other public institutions. The contract for institutional food was usually offered to the lowest bidder, and it was an economic impossibility to become the lowest bidder without some swindling.[22] There was a scandal in 1850 when a large number of orphans in Drouitt's Institution for pauper children died, as a result of their oatmeal being padded with barleymeal, which was less nutritious and gave the miserable children vomiting and diarrhoea. The same thing happened in 1852 when the London Poor Law Unions requested estimates for being supplied with oatmeal. Numerous estimates were sent, and one came in a full three shillings a load beneath the next cheapest. How? Because the oatmeal had once again been mixed with inferior barleymeal.

These cases outraged contemporaries. One writer called this barleymeal adulteration "a barefaced and heartless robbery" practised on the poor, adding: "This species of robbing the poor merits the unmitigated abhorrence of every man who possesses the least particle of either honesty or humanity." But these scandals were just the tip of the iceberg; adulteration was a routine feature of almost all transactions between food-sellers and their impoverished as well as their parsimonious customers. The real mystery was why this incessant robbing of the poor was allowed to continue. Why did England in 1850, in some ways the most industrious and prosperous country in the world, and arguably the most self-satisfied, continue to feed its population so badly?

The Swindlers of England

Both inside and outside Britain, people noticed that adulteration in the country was unusually acute. "There is not a country in the world where commercial roguery is so generally and successfully practised as in Great Britain," complained an anonymous English critic of adulteration in 1855.[23] Several French writers reported the situation in London as worse than that in Continental Europe.[24] Eliza Acton complained that a country possessing the "agricultural and commercial advantages" of England ought to have been famous for "the purity and excellence of its bread," instead of which, it was noted "both at home and abroad, for its want of genuineness and the faulty

PUNCH, OR THE LONDON CHARIVARI.—August 4, 1855.

THE USE OF ADULTERATION.

Little Girl. "IF YOU PLEASE, SIR, MOTHER SAYS, WILL YOU LET HER HAVE A QUARTER OF A POUND OF YOUR BEST TEA TO KILL THE RATS WITH, AND A OUNCE OF CHOCOLATE AS WOULD GET RID OF THE BLACK BEADLES?"

A Victorian cartoon showing a little girl purchasing poisons from a grocer.

mode of its preparation."[25] There was a remarkable consensus on the causes of Britain's problem: the increasing dominance of a laissez-faire mentality, coupled with an absence of adequate laws or proper enforcement regarding the quaity of food.

A French chemist, Jean-Baptiste Alphonse Chevallier (1793–1879), saw a direct relationship between the fact that Britain was "the country *par excellence* of commercial liberty" and the fact that British sweet-makers got away with colouring their candies with copper and verdigris.[26] It would be hard to disagree with this. Except when it affected Treasury revenue—as in the case of sugar, tea, and coffee—the British state did not at this time see it as its business to interfere in the selling of food. Peter L. Simmonds, editor of the *Journal of Commerce*, noted that the Treasury sanctioned "admixtures and adulterations in a variety of instances, as in the case of chicory and coffee, cassia and cinnamon, wild and cultivated nutmegs," but was "most virtuously indignant at adulterated tea, tobacco, snuff and other heavily taxed items."[27]

The British approach was not purely cynical, though. It was also based on a widely shared conviction that there really was no alternative. Supporters of laissez-faire economics (of whom Britain had more than its fair share) had convinced themselves that doing nothing was the best thing to do. The market was god, and many believed that, through some magical process of equilibrium, the market would provide. In the early nineteenth century, many of the old monopolies and tariffs were being done away with in the name of progress. In 1815, the Assize of Bread and Ale was finally abolished as archaic. Without the old guild structure in place, the old assize had become impossible to administer. But instead of replacing it with a more modern form of regulation, Parliament decided it would be best to leave well alone. What happened between a man and his baker was no business of the state; free trade was best. The committee that had been called to comment on the assize decided that "more benefit is likely to result from the effects of a free competition . . . than can be expected to result from any regulations or restrictions under which the bakers could possibly be placed."[28]

The effect of abolition, however, was to transform baking into "one of the most depressed, overcrowded and unremunerative trades of the day."[29] Under the terms of the assize, the price of bread had always been fixed, which guaranteed a certain security to the profession of baker. The relaxing of the law meant that thousands of new bakers set up shop, and all of them sought to undercut one another. Eliza Acton noted that in 1851, the number of bakers in Paris was limited to 601, which meant that they were all assured of selling plenty of bread and enjoying a certain dignity, making a fine product and being paid a decent amount for it, whereas Britain's free trade had pushed the official number of London bakers to 2,286 (the unofficial figure may have been as high as 50,000).[30] These bakers may have had commercial liberty, but they had no peace of mind. To make a loaf that they could sell at a price at which they could find buyers for it, they were forced to reduce the quality of ingredients to a minimum. Under these conditions, honesty was suicidal. A witness to the Committee on Journeymen Bakers commented that "They [the bakers] only exist now by first defrauding the public, and next getting eighteen hours work out of the men for the next twelve hours."[31]

A French chemist who had established himself in London, Alphonse Normandy, observed rather smugly how different things were across the channel. As a Frenchman, he had often been amazed at how shameless were the adulterations that took place in England, where manufacturers had so little to fear from the law. Once, he confronted a baker in Islington with a loaf of bread glistening with alum crystals, only to be insulted with "a very offensive expression about my eyes." By contrast, it was very hard, he said, to adulterate bread in France:

> When bread is adulterated in France, which occasionally is the case, the baker is at once summoned before the police correctional; if it is the first offence, he is fined, or if the offence has been very gross he is shut up for a week or ten days, or something of that kind; and if the offence is repeated, he is prevented from establishing himself again as a master baker; he can work as a journeyman baker, but he can no longer establish himself as a master baker, or the sentence is placarded about town; in fact he is a ruined man.[32]

Food frauds were not unknown on the Continent. In 1844, a whole Belgian family was poisoned after a baker mixed too much copper sulphate in his dough, an additive used to improve bad flour, which one chemist called "hateful fraud."[33] However, such crimes were dealt with far more strictly by the law and its agents in France and Belgium than they were in Britain, where a mere "pecuniary loss" or fine was all that the fraudsters had to fear.[34]

Under Prussian municipal law, for example, it was explicitly stated that "no person shall knowingly sell or communicate to another for their use, articles of food and drink, which possess properties prejudicial to health, under a penalty of fine or bodily punishment ... those who are found guilty of knowingly selling victuals which are damaged or spoiled or mixed with deleterious additives shall be rendered incapable for ever of carrying on the same branch of business," and the fraudulent food would be seized and either destroyed or, if not inedible, "confiscated for the benefit of the poor."[35] This law was in the medieval tradition of laws against swindling. Modern commercial Britain, by contrast, seemed to think it could do without such laws.

John Mitchell, a chemist, expressed amazement in 1848 that England "is about the only nation that has no laws, or no effective laws, for the protection of the public against the adulteration of food."[36] This anomaly was puzzling. By the nineteenth century, the guild system that had done so much to protect food quality was gone from France as well as England (it was finally abolished by the Jacobins). Paris, like London, was a modern industrial city, with consumers distanced from the producers of most of their food. So why was French food not falsified to the same extent? The difference was that the French state had continued the guild's role of protecting the citizen-consumer from bad food. Napoleon's Civil Code ruled that "no person could exercise the trade of a baker without the permission of the prefect of Paris; and no baker could quit his business without having given six months previous notice."[37] The same mentality continued under successive governments. In 1817, after Napoleon had been supplanted by the Bourbon monarchy, police were ordered to maintain an active surveillance of bakeries.

Under the laissez-faire British government, there was a different conception of responsibility from that of Continental governments. In France, the responsibility for producing good food lay with the producer; the state would police their activities and, if they should fail, would punish them for neglecting the interests of its citizens. By contrast, the British government—except in extreme cases—placed most of the responsibility with the individual consumer. It would be unfair, or so the thinking went, and contrary to liberty, to interfere with the shopkeeper's right to make money. In the 1840s, a patent was granted for a machine designed for making fake coffee beans out of chicory, using the same technology that went into manufacturing bullets.[38] This machine was clearly designed for the purposes of swindling, and yet the government allowed it. A machine for forging money would never have been licensed, so why this? As one consumer complained, the British system of government was weighted against the consumer in favour of the adulterator: "Any attempt at fraud on the part of the customer is punished by law, and above all, is easily detected. The bad shilling is at once recognized and nailed to the counter; but the poisonous adulterations practised upon food remain undiscovered, until their effects are shown in the indisposition or perhaps serious illness of the consumer."[39]

Eliza Acton thought that many in Britain chose to "shut their eyes" to the true state of food and drink. "They do not wish to be disturbed in their belief that it is all that it ought to be."[40] In many cases, shutting their eyes to the problem was all that British consumers could do. If adulterated food was the only food on offer, it made psychological sense to pretend to yourself that it was all right, otherwise you would never have eaten anything. The British system of trade in food at this time depended on this enforced suspension of disbelief: the ability of consumers to lie to themselves about what they ate. For this reason, the necessary reform could not come from the consumers themselves, it had to come from government. Yet the British government was equally good at shutting its eyes to the consequences of swindling.

Mid-nineteenth-century France was a very different place. Like several other European countries, France had a board of health—the Conseil de Salubrité—whose job it was to watch over anything that might harm public health, including adulterated food. In Paris, there were seven expert members on the Conseil "who have the surveillance of markets, factories, places of public amusement, bakeries, shambles, meat, medicines, etc."[41] Then, in 1851, France adopted its first generalized law against food adulteration, which put it on a par with financial fraud. French law took the side of the consumer, seeing adulteration as an attack on private property. Unlike the British establishment, which feared that too much food regulation would damage economic life, the French attitude was that "protecting the quality of edible goods encourages the growth of productivity and preserves the reputation of national products." Under the new law the "moral character" of the offence of falsifying food was viewed as serious; it was deemed a grave offence (*délit*) rather than a minor economic infringement (*contravention*).[42]

These national differences were not just abstract. The differing attitudes toward food law in Britain and France yielded substantial disparities in the degree to which the food sold in shops was falsified. Coloured confectionery was perhaps the clearest and most chilling example. In France, Switzerland, and Belgium, sellers of confectionery were held responsible if what they sold turned out to be poisonous.[43] It was expressly forbidden in France to make use of any mineral substance for colouring sweets, lozenges, sweetmeats, pastries, or liqueurs. If sweets were coloured it must be with "safe," mostly vegetable dyes, such as saffron for yellow sweets or cochineal for red ones. It was also forbidden to wrap sweetmeats in paper that had been glazed and coloured with mineral substances. All confectioners and grocers were obliged by law to have their name and address printed in the wrapping paper.[44] If any sweet should be coloured with a poison, the vendor was personally responsible. "But in England," one consumer protested, "the centre of civilization as we are so fond of calling it—poison is openly vended in the streets, shop-windows are filled with it."[45]

This sounds like scaremongering, but it was a statement of fact. In 1831, a Dr. O'Shaughnessy, working on behalf of the medical journal the *Lancet*, toured the streets of London collecting numerous samples of sweets, bonbons, and sugarplums and submitting them to chemical analysis. He found that parents who bought these treats for their darling children were dicing with death. Of the samples collected, the red ones were often coloured with lead or mercury; the green sweets, with copper-based dyes; and the yellow, with gamboge, a purgative resin-based dye from the Far East now used to colour Buddhist robes, or, more perilously, yellow chromate or chrome yellow of lead. How common were these poisonous dyes? If O'Shaughnessy's evidence was typical, they were widespread. Out of ten red samples, two contained harmless cochineal (from crushed insects, which most people can tolerate, though a few people are allergic to it), two contained semiharmless "vegetable lakes of aluminium and lime" (azo dyes such as carmoisine, which are now thought to cause temper tantrums and hyperactivity in some children), and six contained either "red oxide of lead," "red sulphuret of mercury" (vermilion or mercuric sulphide), or lead chromate, any one of which could give a child a nasty dose of heavy metal poisoning.[46] Despite O'Shaughnessy's work, the sale of poisoned sweets continued unabated over the next two decades.

The British famously had a greater fondness for sugar than the French, and it manifested itself in the yearning of the British child for fantastically coloured sweets. The pick 'n' mix sweets of the modern British child—the pink shrimps, fried eggs, cola bottles, and so on—had equivalents in the 1840s. The Victorian confectioner sold ginger pearls and sugar dragees, yellow rock and multicoloured hundreds and thousands, clove sticks and peppermint pipes, coconut candy made from sticky brown sugar, strawberry sweets and apple sweets, sugar oranges and sugar lemons, and all had to come in a blinding array of colours if they were to sell. One contemporary noted that

> In the large and frequented thoroughfares, such as Tottenham Court Road, Houndsditch or High Street, Whitechapel, these establishments are made as showy as possible; they burn a vast amount of gas, and have their windows filled with sugar compounds, many of which have been moulded into fanciful and highly coloured forms. Sometimes the image represented is a mutton chop, or rasher of bacon; onions and potatoes are very popular; and eggs and oysters, dogs, shoulders of mutton, pears and mackerel are also much esteemed by the youthful customer.[47]

A candy mutton chop is a pretty frightful idea, particularly if you consider the fact that the red grain of the meat was almost certainly painted on using lead-based colouring.

Unsurprisingly, there were frequent reports in British newspapers of poisonings from sweets. In September 1847, three adults and eight children were taken to Marylebone Workhouse "having been seized with vomiting and retching after eating some coloured confectionery."[48] The following year, the *Northampton Herald* reported that several people had been poisoned at a public dinner on account of some green sugared cucumber ornaments used to decorate a blancmange. One man died as a result. The year after that, some children in Marlborough became alarmingly sick after eating "a green flower made in sweet paste in imitation of a fuchsia," which had been used to decorate a "magnificent cake."[49] One French scientist commented with Gallic bemusement that every year in England a number of children died as a result of eating toxic sweets.[50] Something had clearly gone wrong with a food culture that took such trouble to make such grotesquely inedible foods.

It was not only the French who noticed, however. An anonymous English writer complained that such things were bound to happen for as long as the government maintained its blind attachment to the beneficent workings of the free market: "in England, the only protection for the public, under the present laissez-faire system of government, is that afforded by *publications* in which the means of detecting adulterations are clearly indicated."[51] Even publications did not necessarily provide protection though. Just because the news papers had exposed one batch of green copper sweetmeats as toxic, it was no guarantee against other kinds of toxic green sweets being made and sold by other confectioners. By 1850, there was plenty of publicity in Britain about adulterated food. Most of it, though, was too vague and too unscientific to do much good.

Publicity vs. Science

These days, the newspapers seem so full of contradictory food scares that it becomes difficult to see how anyone has ever managed to eat anything without their health suffering. One week it is "Eat more oily fish or miss out on valuable omega 3s!" and the next it is "Too much oily fish will give you mercury poisoning!" Before long you glaze over when reading these stories. You develop food-scare fatigue. Because newspapers deal in fear, and most of us are not toxicologists who can unravel the ins and outs of a story, it becomes hard to distinguish alarmism from truth. This problem is not new.

In 1830, ten years after Accum's *Treatise*, another book appeared denouncing the whole spectrum of food and drink in Britain as fraudulent. It was entitled *Deadly Adulteration and Slow Poisoning; or Disease and Death in the Pot and the Bottle*. Unlike Accum's book, it was not signed, but written anonymously "by an enemy of fraud and villainy." Unlike Accum, this anonymous author was clearly not a scientist. Instead of hard evidence, he indulges in scaremongering and wild accusations, lashing out vaguely against "charlatans and nostrum-mongers" and the "pernicious system of fraud" that physicians as well as food-sellers were guilty of.[52] Like many modern-day conspiracy theorists, the author mixes up truth and fantasy until you can hardly be sure which is which. He writes—accurately—that many potato sellers tried to increase the weight of their "murphies" by soaking them in water.[53] He complains—truthfully—about the deadly cosmetics that so many ladies slapped on their faces. But then he writes—absurdly—that drinking too many spirits will make a person die of "spontaneous combustion" because the spirits have so transformed the human body that it must suffer a kind of "supernatural" punishment.[54] The trouble with this kind of scaremongering was that its style made it hard for the reader to believe even what was true. The *Lancet*, ever sensible, characterized the author as a "well-meaning individual, but of that class of exaggerating alarmist" who adopt a "tone of half-mad honesty."[55]

Far from checking the extent of adulteration, such tracts could encourage it by making swindling seem normal; if swindling was everywhere, what was the point of being honest? With its scattergun approach, accusing all food producers in general rather than naming specific guilty parties as Accum had done, such publicity was of little value to the consumer. It was scant help, when buying groceries from your local shops, to know that grocers in generally tampered with food; what you needed to know was whether this particular grocer was one of the demons, or an honest exception. What's more, by sensationalizing the problem, these tracts could leave the rational gentlemen of the British establishment feeling that it was best to let the matter alone. Even those Victorians who were staunchly anti-adulteration might feel that while "the press has literally groaned with the efforts of sensational writers on this subject . . . it has often been grossly exaggerated."[56] Scaremongering only bolstered the position of the laissez-faire dogmatists that the best thing to do was to do nothing. And while this attitude dominated, adulteration became ever more prevalent.

By 1848, John Mitchell could write about adulteration in Britain as a "growing evil" that had only got worse in the thirty years since Accum, with some "extraordinary and newer frauds" taking their place alongside the tried and tested ones. Mitchell was the author of a *Treatise on the Falsification of Food and the Chemical Means Employed to Detect Them*, which took up where Accum left off. Unlike the scaremongers, Mitchell backed up his claims with evidence, analysing many foods himself or else relying on other scientific authorities. It is therefore telling that he repeated most of Accum's accusations—red lead in Double Gloucester cheese; alum in bread; plaster of Paris, pea, and bean flour in flour; no end of poison in confectionery—and added some more of his own: milk adulterated with water, flour, milk of almonds, gum, chalk, and turmeric; "so-called finest chocolate" adulterated with flour, potato starch, and—how revolting this must have been—clarified mutton suet.[57] Accum had described a scenario where developments in chemistry contributed both to new techniques in swindling and to new means to fight it. This had only continued, in Mitchell's view. "As chemistry advanced, it unfolded new secrets, and opening on the one hand more decided and unequivocal tests for the adulterations, it at the same time gave a larger scope for the adulterators." The situation was so bad, he believed, that "nearly all the substances used . . . are adulterated, debased or badly manufactured."[58]

This was despite the fact that since the beginning of the nineteenth century, as well as refining of old techniques, chemists had developed numerous new techniques for analysing food and drink. Mitchell lists countless new gadgets and new methods: Schuster's alkalimeter, a teardrop-shaped vessel for testing the pH value of beer

and wine; Gay-Lussac's alcoholometer, a more accurate tool for measuring the alcohol content of drinks than the old hydrometers for
measuring specific gravity mentioned by Accum; the Pesier natrom-
eter for measuring the purity of potassium salts; and the chlorometer
and acetimeter as used by Chevalier. From the 1820s onwards, Gay-
Lussac developed volumetric analysis, whereby the nature of liquids
was gauged through the way their volume fluctuated when certain
reagents were added. The 1830s saw the birth of organic chemistry—
the chemistry of animal and vegetable substances, as opposed to
minerals—as a distinct branch of the subject, enabling many foods to
be analysed far more accurately, and broken down into their distinct
components. Mitchell cites a French chemist called Jean-Baptiste
Dumas, who analysed flour into albumen, fibrine, caseine, glutine,
starch, and glucose.[59] There were also new developments in colo-
rimetry, where suspicious ingredients could be indentified by com-
paring their colour when dissolved in solutions with the colours of
known chemicals. This was a useful test for finding out, for example,
if saffron was pure, as saffron solution is a very distinctive yellow.
Fluorimetry—testing the fluorescence of different solutions—was
likewise useful for identifying the purity of mustard as against turmeric, since turmeric is strongly fluorescent, while mustard is not
fluorescent at all.[60]

However, hardly any of this chemical analysis of food was taking
place in Britain. English analytical chemists were in short supply,
whereas France—and later, Germany—abounded with them: Gay-
Lussac, Dumas (1800–1884), Descroizilles, Vauquelin, Pelouze, Pé-
ligot.[61] Thus, while the French had in place both the governmental
structure and the body of scientific experts needed to deal with the
problem of food swindling, Britain had neither. Moreover, the new
chemistry, though brilliant at certain tasks, wasn't perfect. If you
wanted to test the alcohol content of brandy or to determine whether
certain mineral additives had been included in food, you had a better
chance of success now than ever before. If, however, you wanted to
know if milk was pure, you were on shakier ground.

Mitchell admitted the difficulty. Water, he said, was "the substance
employed most generally" to adulterate milk. This was "very difficult

to detect" because "the density of pure milk is variable." So the usual chemical tests done to test the density of a substance could not establish with absolute certainty whether milk had been watered down, assuming the percentage of water used was fairly small. There were similar problems in detecting the near-universal mixture of expensive coffee with cheap chicory. If you had nothing but chicory on your hands, you could do a simple chemical test: nitrate of silver, says Mitchell, gives no precipitate with chicory, whereas it does with coffee. But this test would be useless to detect a mixture of chicory and coffee; the precipitate would still ensue, on account of the presence of coffee; the test would only come out completely negative if a shopkeeper had been brazen enough to sell pure chicory as pure coffee. This left great uncertainty. Indeed, in 1847, Sir Charles Wood, the laissez-faire chancellor of the Exchequer (mainly remembered now for keeping poor relief as low as possible), told the Commons that "three distinguished chemists" reported to him that "neither by chemistry nor in any other way can the admixture of coffee with chicory be detected."[62] The inference was that if it couldn't be detected, then nothing could, or should, be done about it.

Wood turned out to be wrong, in every possible way. This mixture—along with other food mixtures—*could* be detected with absolute certainty, and the method for doing so was about to be discovered. The problem was that food scientists had been looking in the wrong direction. Chemical analysis alone was not going to crack the problem. What was needed was a scientific instrument that had, bafflingly, been more or less neglected by food analysts until the 1850s: the microscope. The tide turned decisively against swindling in Britain only after food was put under the critical gaze of the microscope, and not by a chemist but by a doctor, the great and faintly absurd Dr. Arthur Hill Hassall, "the apostle of antiadulteration."[63]

Food under the Microscope: Arthur Hill Hassall

"In those days," recalled Arthur Hill Hassall, looking back over his long and busy life to his youth in the 1840s, "people often said to me, 'Ah! The microscope is all very well as an amusement, but of what

119

practical use is it in life?,' these people little dreaming of the many and vastly important facts which in the future were to be brought to light by its instrumentality." In the twenty-first century, microscopy

Arthur Hill Hassall (1817–94), the scourge of Victorian swindlers. Hassall was about forty-five when this mezzotint was made. In the background is a statuette presented to him in 1865, depicting fraud (the toad) being speared by science.

is fundamental to almost all chemical research. But in the 1840s, the microscope was treated by chemists as little more than a "scientific toy"—entertaining, rather than useful. The compound microscope had been used by botanists and biologists since the seventeenth century and had plumbed such mysteries as the sex of bees, yet "Hassall was the first to realize the full and systematic application of microscopy to the detection of adulteration."[64] Without the microscope,

Hassall wrote, the "multitudinous adulterations practised on nearly every article of consumption" could "never have been discovered and exposed."[65] Chemical analysis was useful as far as it went—for detecting copper in pickles, say, or lead in cayenne—but it was "powerless" to discover "organic admixtures" of foods. It was the microscope, said Hassall proudly, that was "the great and chief means of detection"; and Hassall himself was the great and chief wielder of it.

In some ways, Hassall was an unlikely social campaigner. The son of an army doctor, he was a stiff, neurotic figure, notwithstanding his luxuriant sideburns (which he accessorized, in later life, with an equally luxuriant moustache), possessing a long nose and a pinched mouth, a man whose emotions were decidedly closed off; he managed to write a 166-page autobiography without mentioning that he had been married—twice—while dwelling at length on his boyhood reactions to eating a blood pudding. But Hassall's repressed nature was allied with a fierce moral indignation, which made him recoil instinctively from fraud and lies. In 1850, he was a thirty-three-year-old doctor who had already retired from his Notting Hill practice on grounds of ill health. He had found his practice "somewhat harassing"—as he seems to have found much of life—and then, after walking home late from the theatre one night, he got wet through and developed pleurisy, from which he never fully recovered. As a result, he and his first wife left the harassing patients of Notting Hill and moved to a house in St. James's, where by way of a hobby he fitted himself up with a modest laboratory for "chemical and microscopial research." The history of British food might have been very different if Hassall had chosen stamp collecting instead.

During this premature retirement, Hassall was in the habit of energetic "journeyings" through the streets of London. He couldn't help noticing the "various articles of consumption" on display in shop windows and "the explanatory placards exhibited." It was Hassall's instinct that "there was much amiss in the appearance of some of the articles and statements made in respect of them"—in other words, he didn't believe the hype. Meanwhile, he kept noticing in the newspapers "frequent complaints of the bad quality of the ground coffee sold and many doubts expressed as to its genuineness." Being at a loose end, Hassall decided he would "look into the matter myself."[66] He realized that, before he could determine whether coffee in the shops was adulterated, he would have to know what both coffee and chicory looked like under the microscope in their pure forms. So he set to work.

> I commenced by examining microscopically sections of the whole coffee berry, unroasted and roasted, and then the roasted berry after being ground. In the same way I examined the raw and roasted chicory root. I had now some valuable data to proceed upon. I found that the roasting and partial charring and blackening by no means destroyed the beautiful minute structures and tissues entering into the composition of the coffee berry and chicory root.[67]

Coffee adulterated with wheat and chicory viewed through Hassall's microscope.

Looking at Hassall's illustrations, you can see exactly what he meant. Under the microscope, coffee and chicory are chalk and cheese. Pure ground coffee looks like shards of honeycomb, whereas pure ground chicory looks like slices of squidgy cucumber, on account of its milky "utricles" or cells.[68] This was an immensely useful discovery, because Hassall could now tell, beyond a shadow of a doubt, whether any given sample of coffee had been adulterated. His microscope could detect when even a tiny amount of chicory had been added to a sample of coffee, because it wouldn't take long before a single chicory utricle would show up.

His next move was to go into London shops, buying samples of coffee and analysing them. His results showed that nearly all the coffee was "adulterated most extensively in a variety of ways," and "some consisted of little else than chicory." Other samples when put under the microscope exposed "considerable amounts" of roasted wheat, rye, beans, and burnt sugar, and "these spurious admixtures were sold under the most grandiloquent names and with statements absolutely false."[69] (He does not seem to have found any liver, but then, this was added only to ready-prepared coffee, not mixed into the ground stuff.) Hassall wrote his findings up, and on 2 August 1850 read a paper *On the Adulteration of Coffee* before the Botanical Society of London.[70] The effect was immediate. The following week, *The Times* carried a leading article on Hassall's discovery. Suddenly, his "retirement" from medical practice was becoming rather eventful.

"Thus encouraged," Hassall set to work on other foods.[71] He chose brown sugar, which had a very different structure from coffee, being "crystalline" rather than "organized." Popular opinion believed that grocers were often diluting their sugar with sand to make it go further. A nineteenth-century joke repeated in various forms had a grocer asking his assistant: "Have you watered the treacle and sanded the sugar? Then you may come in to prayers."[72] Hassall's microscope discovered that this particular food scare was not true; he found no evidence of sand in any of the samples he examined, concluding that "the grocer was probably libelled" when he was accused of sanding his sugar. Not that Hassall could recommend eating brown sugar. It may not have contained sand, but it did abound "with living and dead acari, louse-like creatures, in all stages of growth and development," which explained "the malady to which grocer's assistants were specially liable, namely grocer's itch."[73] Until brown sugar could be supplied in a purer form, Hassall advised eating white sugar instead.

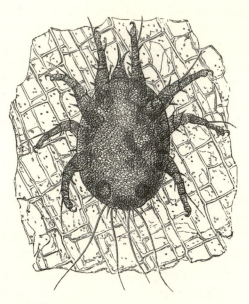

The sugar mite, which Hassall found in brown sugar. The mite was guilty of causing the nasty condition, grocer's itch.

The great advantage that Hassall had over previous antiswindling campaigners was the pitiless accuracy of his microscope. Even so, what was to guarantee that the brief food scandal he had created would not fade, as Accum's had done? What made Hassall's scientific work so lasting was the way it was yoked to a relentless and perfectly executed campaign of publicity, and for this the credit must go not to Hassall himself but to Thomas Wakley, the visionary editor of the *Lancet*.

Naming and Shaming and the Health of a Nation

By the time Hassall's microscope was causing its stir, Thomas Wakley (1795–1862) had been looking for several decades for a way of reforming British food. A "lifelong radical," he had set up the weekly medical paper, the *Lancet*, after his promising medical career was damaged by some macabre personal rumours. In August 1821, Wakley, then a young and recently married surgeon, was assaulted in his own home, and the house burned down. The insurance company accused him of arson, because the house had been heavily insured. Meanwhile, another rumour circulated that Wakley was the surgeon responsible for having beheaded the corpses of five political extremists who had been hanged earlier that year. So far as we can gather, there was no truth in either rumour. In 1821, Wakley successfully sued the insurance company over the arson claim, receiving full compensation. He had been shaken by the experience, however, as had his wife, and he sought a new direction.

THE LATE MR. WAKLEY.—SEE SUPPLEMENT, PAGE 621.

Thomas Wakley (1795–1862), founder and editor of the *Lancet*, which gave Hassall's science the publicity it needed.

The word "lancet" had more than one meaning, as Wakley was aware. Most obviously, it was a medical instrument, a kind of scalpel, but double-edged, which could be used to make incisions, just as the *Lancet* aimed to cut out both nonsense and disease. Less well known was its meaning as an architectural term: a lancet is a Gothic arched window, designed for letting in light. From the very beginning, the *Lancet* had an agenda of enlightenment as well as criticism; Wakley believed passionately in the power of the right kind of publicity to effect good and aimed his paper at the general public rather than a narrow medical audience. His vision of health was social, not just individual. The *Lancet* campaigned for hospitals to make their statistics public and for the medical corporations to become more open, professional, and democratic. It attacked the enemies of health wherever it saw them—quacks, inept doctors, the inadequate poor relief system, the callous system of corporal punishment in the armed forces. Wakley argued for public health in the widest sense—and this necessarily entailed a frontal assault on the evils of adulterated food.

It had been Wakley in 1831 who had hired the young medical graduate W. B. O'Shaughnessy to go out onto the London streets to collect samples of coloured confectionery for analysis. Wakley had read one of the scaremongering anti-adulteration tracts we have already encountered, *Deadly Adulteration and Slow Poisoning*, a book that left Wakley in two minds. His instincts told him that much of what the book said was true, but he disliked its wild tone, unsupported statements, and inaccurate science, which, far from contributing to public health, could only lead to an "epidemic of terror."[74] In Wakley's view, the only way to take on the swindlers was to hold up "individual malefactors" to "public animadversion." "To attack vice in the abstract, without attacking persons, may be safe fighting indeed, but it is fighting with shadows."[75] This was exactly what he and O'Shaughnessy attempted with the *Lancet*'s article on poisoned sweets, which published the names and addresses of the guilty confectioners. So far as it went, it was a hit. O'Shaughnessy's analyses confirmed that the scaremongers were right, and the *Lancet* urged the government that Something Ought to Be Done. As usual, though, nothing was. O'Shaughnessy joined the East India Company and left for India. Wakley employed another scientist, T. H. Henry, as a food analyst, but he didn't discover anything interesting enough to publish in the *Lancet*—whether for lack of talent or because he analysed the wrong things, we don't know.

When finally Hassall published his revolutionary findings about adulterated coffee, Wakley's excitement was palpable. Here, at last, was the scientific tool—the microscope—to power Wakley's publicity machine. He wrote to Hassall at once, with all the conviction of someone who had spent twenty years pondering the question: "You will never effect any lasting good until you are able to publish the names and addresses of the parties of whom the articles were purchased, giving the results of the examination in all cases whether good or bad. Do you think it would be possible to do this without an amount of risk which might be ruinous?"[76] Despite his cautious nature, and despite fearing that he "risked all I possessed, namely, my scientific and professional reputation," Hassall bravely replied: "Yes, I believe it might be done."

The pair wasted no time in setting up the rules by which they would operate. Hassall would—anonymously, at first—write a series of articles for the *Lancet* analysing food and drink samples purchased from shops all over London. These articles would appear frequently and would cover as many different kinds of food and drink as possible. Every article would include the names and addresses of the sellers of adulterated samples. Wakley would bear the expense and the legal risk, and Hassall would do all the spadework, accompanied by a helper, Henry Miller (who also made the beautiful drawings of Hassall's microscopial analysis), so that one of them could make the purchase and the other could be, "if needed, a competent witness" to the sale. On leaving the shop, they would write the name, vendor, date, and cost of the product on the wrapping, plus each of their initials, to ensure that no mistake was made. It was tough work, especially for someone of a frail disposition. The *Lancet* reports appeared initially every week, and then fortnightly, through the years 1851–54. To obtain enough samples, Hassall and Miller had to make frequent "nocturnal excursions" into the seedier parts of London "in all weathers and in all seasons." Often, through "waiting and hanging about," they became "chilled to the bones, not arriving home till near midnight."[77]

It was worth it, though. For the first time, here were facts about adulteration that were statistical rather than anecdotal. Over these four years, Hassall analysed more than 2,500 samples of food embracing "all the principle articles of consumption, both solids and liquids" and found that purity was the exception, adulteration the rule.[78] Earlier writers might say vaguely that cinnamon was "often" or "sometimes" adulterated (with cassia, wheat, mustard husks, and colouring), whereas Hassall could state with absolute certainty that out of nineteen samples of ground cinnamon, only six were genuine; that three consisted of nothing but cassia; that ten were mixed up with bulking agents such as sago, flour, or arrowroot; and that these faked cinnamons were not always cheaper than the real thing, meaning that the public was being consistently cheated in the purchase of cinnamon.[79] Unlike the scaremongers, Hassall was not afraid to say when a food was *not* adulterated. All of the twelve samples of mace that he tested were genuine. Salt, too, was generally pure. On the other hand, he and Miller couldn't find a single sample of unadulterated mustard in the whole of London, no matter what price they paid.

The effect of this relentless, hard-edged publication of the facts of food swindling, week after week, was to shake British public opinion out of its apathy over adulteration. After Wakley's death, Hassall had the good grace to acknowledge that this was largely due to his colleague's "moral courage" and his "bold and unprecedented step" of printing the names and addresses of tradesmen (though in his lifetime it rankled with Wakley that he received so little of the credit and Hassall so much).[80] A contemporary observed that

> A gun suddenly fired into a rookery could not cause greater commotion than did this publication of the names of dishonest tradesmen, nor does the daylight, when you lift a stone, startle ugly and loathsome things more quickly than the pencil of light streaming through a quarter-inch lens, surprised in their native ugliness the thousand and one illegal substances which enter more or less into every article of food which it will pay to adulterate.[81]

Although this work was long overdue, in some ways the time was right. There had been a shift in mood since Accum's time. Hygiene and public health were now properly on the political agenda; the Victorians had become alarmed by their own unhealthiness and wished to do something about it. From 1836 to 1842, Britain had suffered unprecedented epidemics of cholera, typhoid, and influenza. In 1842, Edwin Chadwick (1800–1890) had published his Report on the *Sanitary Condition of the Labouring Population of Great Britain*, exposing the filth and poor drainage suffered by the working classes and linking this to high rates of disease and mortality. The following year, there was a Royal Commission on the health of towns; and in 1848, the government finally caught up with the Continent and set up a General Board of Health, with Chadwick as its chief commissioner. The new mania for public health created unlikely political allies. Evangelical Christians now united with irreligious Benthamites over the attractiveness of sewers. Something else was changing too, as Chadwick observed. Laissez-faire was slowly turning from a term of approval to an insult. For some, it no longer meant freedom, but the selfish, uncaring attitude of "letting mischief work."[82] Trying to improve the nation's health no longer seemed like unnecessary interference, but the only rational—or Christian—thing to do.

By 1850, to be "sanitary" was the height of desirability. This was reflected in the name that Wakley gave to his and Hassall's joint project—the Analytical Sanitary Commission. Every report on adulterated food and drink published by Hassall appeared under this unwieldy title. Before the first report appeared, the *Lancet* made a grand announcement: "uncontaminated air and pure water are now universally regarded as necessary to the maintenance of healthy existence, and to obtain them we have appointed Boards of Health and Commissions of Sewers."[83] Tackling adulteration was the obvious next step. The *Lancet* proposed "to institute an extensive and somewhat vigorous series of investigations into the present condition of the various articles of diet supplied to the inhabitants of this great metropolis and its vicinity."

One of the earlier and more dramatic reports in the *Lancet* was on water itself. Consumers had known for decades that most of what came out of the tap was foul. But they had no way of proving it. Despite the existence of several chemical tests, scientists were almost as powerless. In 1828, a Dr. William Lambe of the Royal College of Physicians wrote that most of the water drunk by ordinary people was deadly; but he could not specify exactly how or why. It was easy enough to tell when water was stagnant or putrid, but much water contamination was imperceptible to the human eye. It takes only an infinitesimal amount of the worst organic contaminants for water to be capable of causing death. Enter Hassall and his microscope. With his usual thoroughness, Hassall collected samples of water from all the main water companies of London: from Chelsea and Lambeth and Vauxhall and Hampstead and East London, and also from some outside London, such as Kent. Under the microscope, Hassall found levels of impurity never before suspected. In his youth, he had done some work on varieties of freshwater algae. The living organisms he found in Thames water were a good deal less charming.

The seepage of sewage into the water supply meant that "drinking" water was teeming with "excessive contamination with organic matter, animal and vegetable, dead and living."[84] The engravings accompanying the article illustrated the contamination in disgusting detail. In 1850, Hassall had published a book with full-colour illustrations of revolting organisms, *A Microscopical Examination of the Water Supplied to the Inhabitants of London and the Suburban Districts Punch* carried a cartoon exaggerating Hassall's illustrations: "a drop of London water," which was teeming with skulls, turtles, humanoid bugs, and tombstones. Actually, this was not so far off the original. Hassall discovered that West Middlesex water, for example, was infected with countless tiny crablike animals (*entomostroceae*), together with algae and fungi from sewage, despite the claim of the water company that their water was "bright and pure at all seasons."[85] As Hassall pointed out, water contamination mattered not just in its own right, but because of its use in other adulterations, "since in milk, beer and spirits the chief adulterant employed in London was Thames or some other impure water."[86] Even if you were canny enough to avoid drinking plain water, you might still be infected by drinking watered-down beer.

Needless to say, the water companies hated being named and shamed. In response to the *Lancet*'s allegations, the Southwark and Vauxhall Water Company protested that it had a full filtration system and that "the company receives no complaints." To which the

A DROP OF LONDON WATER.

A *Punch* cartoon mocking the deplorable state of London water, based on Hassall's microscopic analysis of Thames River water.

Lancet replied: "Complaints indeed! Where is the utility of address-ing complaints to monopolist directors who, in reply, will probably intimate their readiness to cut off the supply altogether!" Besides, whether the company received complaints or not, the microscope proved that Southwark and Vauxhall water was "with one excep-tion, the worst water in London"—swarming with disease-carrying organisms. In 1851, the government held an inquiry against the water companies, at which poor nervous Hassall was obliged to be a witness. The clever barristers for the water companies did their best to humiliate him, preying on his sensitive nature. One of them shouted under their breath, "That humbug, Dr Hassall," a puerile trick to unnerve him. It worked. Forty years later, he still smarted at this slur. Another cast doubt on whether the bottles in which he had collected the water were clean. He assured the court they were. What, had he washed them himself? Hassall replied that he had. To which the barrister triumphantly said: "You are a bottle washer then!" [87]

The lawyers could tease Hassall all they liked. They could make him anxious—this was pathetically easy to do—but they couldn't change the hard facts of his microscope. Hassall noted proudly that during the whole period of the naming and shaming campaign from 1851 to 1854, though "a few lawyers' letters were received and in one or two cases actions were commenced . . . only one went as far as the delivery of the declaration."[88] This, he said, a touch boastfully, was testimony to the "remarkable accuracy" of his reports. The more the swindlers wriggled, the more absurd he could make them look, and they knew it. One of the cleverest aspects of the *Lancet*'s campaign was the way it used the grand lies of commercial advertising against the very products they were advertising.

Advertising and Legislation

Wouldn't it be refreshing if, for every advertisement, there was a counteradvertisement attached? "Contains calcium, good for children's bones," the packaging of a processed cheesy snack might say. But underneath, in big letters, it would read, "Also contains sizeable amounts of saturated fat, salt and colouring, which will do them no good at all. And by the way, there is calcium in *all cheese*, not just in this sorry imitation." This will never happen. The closest anyone has ever got to this fantasy was probably Hassall, who punctured the falsehoods of food marketing with devastating understatement.

Hassall lamented the fact that there was no legal requirement for food packets to state their true ingredients. For the purposes of his rhetoric in his *Lancet* reports, however, this failure of the law was very useful. Hassall's technique was simple. He would quote in full the fanciful claims of the manufacturers, before revealing the true contents of the packet. Again and again, Hassall would cite the assertion of the seller that their product was "genuine," only to show that it was highly adulterated. "Genuine mustard," "genuine cayenne," and "genuine arrow-root" were anything but. "Finest WHITE PEPPER" said a package bought from Wm. Bowley of 110 Tottenham Court Road. Hassall noted that when he visited the shop, this shopman had particularly directed him to choose "the article as one of great superiority and undoubted purity," whereas his own judgement was: "adulterated, consisting of finely-ground black pepper, and a very large quantity of wheat flour."

Hassall was quite explicit about what he was doing. He recognized that he was living in the first great age of advertising and saw it as his job to set the record straight.

> Tradesmen now resort to the press extensively. They are authors and puffers on an extensive scale; witness the numerous handbills, circulars and advertisements, used to announce the various articles of food and drink. It is but right, therefore, that the press should make an exposure of the adulterations perpetrated, and thus supply the antidote as well as the bane.[89]

Some of the most outrageous advertisements made a grand play of how pure the food was. At James Robinson's coffee shop at 156 Bishopsgate Street, there was a placard of gigantic proportions.

> GENUINE COFFEE.
> No adulteration.
> We conceive it is our duty to caution our friends and the public against the present unjust and iniquitous system pursued by grocers in adulterating their coffee with
> Roasted beans,
> Dog biscuit,
> Chicory, and tan.
> Our advice to purchasers of coffee is, to buy it in the berry, and grind it yourselves; if you cannot do this, purchase it of respectable men only: pay a fair and honourable price for it; you may then depend upon a GOOD and GENUINE article.

The implication was that James Robinson himself was "respectable." In fact, Hassall revealed, his coffee was *Adulterated—with a very large quantity of chicory.*"[90]

The more insubstantial the food, the more exaggerated were the blurbs. At the time of the *Lancet* reports, there was a craze for farinaceous foods, special curative foods for invalids, powders to be mixed up with warm water or milk into a magical gruel that promised to answer all the niggling stomach worries of the Victorians. Because these foods could sell for a high price, there were many of them on the market, and because there were many of them on the market, they all made absurdly inflated claims for their own efficacy, insisting that all other farinaceous foods were nothing but snake oil, and that they alone could be trusted.[91] One of the most prominent brands, Warton's Ervalenta, insisted that

> This agreeable, nutritious, farinaceous food radically cures habitual constipation (costiveness), indigestion, piles, and all diseases originating in a disordered state of the bowels and digestive organs, which it speedily restores to their natural vigour and action, *without the aid of medicine*, or any other artificial means ... The invaluable properties and extraordinary efficacy of this eminently curative dietetic have been acknowledged by the first physicians and analytical chemists of the day.[92]

Warton's Ervalenta was pricey stuff. A 1lb canister sold for 2s. 9d. The recommended portion size was 2 ounces, yielding eight portions per canister at just over 4 pence each. Compare this with the cost of bread in 1850, which was about 1.8 pence per pound.[93] In addition, you were advised to purchase a special syrup to take alongside the Ervalenta, called "Warton's Melasse," costing a further shilling a bottle. If the testimonials on the handbill were true, maybe it was worth the high price. Only Ervalenta, it insisted, could restore you to perfect health. It made a great point of warning the consumer to avoid its main rival, "that vile and spurious article called *Re*valenta," which is not, like Ervalenta, "an agreeable and pleasant food, but a nauseous and viscous preparation, more adapted to pig-wash." It also noted that "Persons having mistaken Lentil Flour for 'Warton's Ervalenta', Warton & Co. inform the public that it is a quite different article."

Not so different, as it turned out. Under Hassall's microscope, Warton's overpriced Ervalenta was nothing but ground French lentils, fragments of husk, and starch granules from a substance resembling Indian corn. The special "Melasse" syrup was bog-standard molasses; the only special things were the price and the pompous packaging.

Hassall was especially enraged by the way that advertisers assumed the language of anti-adulteration. The worst swindlers were those who spent the most time attacking swindling. Warton's main rival, Du Barry & Co., who marketed a farinaceous food called Revalenta Arabica, issued a handbill attacking "FIFTY DIFFERENT GANGS OF SWINDLERS" who made a living from selling "trashy compounds of peas, beans, lentils, Indian [corn] and oatmeal." Under the heading "CRUEL DECEPTIONS ON INVALIDS EXPOSED," Du Barry stated, alarmingly, that the health of many invalids had been "fearfully impaired by spurious compounds of peas, beans, lentil, Indian and oat meal, palmed off upon them under closely similar names, such as Ervalenta, Arabica Food, Lentil Powder, Patent Flour of Lentils etc."[94] As for Du Barry's own product, "this light, delicious breakfast farina," it claimed to have even more health-giving properties than Warton's Ervalenta. As well as banishing all stomach complaints, it also removed

> palpitations to the heart, nervous headache, deafness, noises in the head and ears, pains in almost every part of the body, chronic inflammation and ulceration of the stomach, eruptions on the skin, scrofula, consumption, dropsy, rheumatism, gout, nauseas and vomiting during pregnancy, after eating, or at sea; low spirits, spleen, general debility, paralysis, cough, asthma, inquietude, sleeplessness, involuntary blushing, tremors, dislike to society, unfitness for study, delusion, loss of memory, vertigo, blood to the head, exhaustion, melancholy, groundless fear, indecision, wretchedness, thoughts of self-destruction etc.[95]

As you read down the list—a breakfast which can cure indecision!—you begin to wonder if the copywriter is sharing a joke with the consumer. A sick joke, if so. Another striking thing about the handbill was the way it took particular pains to distance Revalenta from any association with lentils. Lentils, the writer insisted, were difficult to digest and caused nervous complaints. They could make a person "very ill." Therefore it could not be *in any sense* true that the Du Barry company was supplied with lentil flour from a certain Mr. Nevill, as this Mr. Nevill had recently said. Du Barry *certainly did not* pay Mr. Nevill "an annual sum" for fabricating the Revalenta Arabica. "So much for Mr Nevill's insane fabrications!"[96] Revalenta, Du Barry maintained, was not made from lentils—perish the thought!—but from "the root of an African plant, somewhat similar to honeysuckle."

After all this hyperbolic nonsense, it is deeply satisfying to read Hassall's analysis of Du Barry's Revalenta, which he purchased from a shop on Oxford Street. Revalenta is nothing but a mixture of lentil and barley-meal. Hassall cannot resist concluding that, since Du Barry has condemned the nutritive properties of lentils, it has, by implication, condemned itself.[97]

In the battle between Hassall and the swindling advertisers, the first round undoubtedly went to Hassall. In 1855, a Parliamentary Committee was set up to look into the adulteration of food and drink, calling Hassall as its first witness. Since the first *Lancet* report in 1851, the cause of anti-adulteration had attracted new campaigners. John Postgate (1820–91) was a Birmingham surgeon who knew all too well the truth of Hassall's reports, having spent his youth working as a grocer's boy, where he had been schooled in swindling. As an adult, Postgate poured everything into fighting adulterated food, even taking money from his own family to print pamphlets on the subject. Postgate enlisted two radical Birmingham MPs, George Muntz (1794–1857) and William Scholefield (1809–67), to the cause. It was Scholefield who headed the Parliamentary Committee.

In front of this body of mainly Radical and Liberal politicians, Hassall was given free rein to repeat most of his main allegations against contemporary shopkeepers—reading out a full list of all the adulterating substances he was aware of and elaborating their ways and means—but this time in a setting where he knew his words might influence law. Now he could wreak his revenge on all the swindlers of England. He even brought in some coloured confectionery and cakes to show the committee, so that they could see with their own eyes "how coarsely and grossly they are coloured"—with chromate of lead, red lead, Prussian blue, arsenite of copper. After displaying these gruesome sweetmeats, Hassall informed the committee that "there is sufficient in one of those cakes to produce some temporary derangement; and if a child were to be in the habit of easting two or three of those cakes, it might injure it very seriously."[98] The committee of bigwigs was evidently impressed by his dramatic evidence and asked Hassall a leading question, revealing their sympathy to his cause. "In the existing state of society do you think that *caveat emptor* should be changed into *caveat venditor*?"—in other words, should "buyer beware" become "seller beware"? Hassall did not hesitate to reply, simply, "Yes."[99]

After questioning numerous other witnesses—scientists, MPs, apothecaries, shopkeepers, and commercial men—the committee reported that "We cannot avoid the conclusion that adulteration widely prevails."[100] In 1860, Britain finally passed its first general Adulteration Act, which owed its existence largely to the publicity of Hassall and the *Lancet*, coupled with the political activism of Postgate and Scholefield. Now, at last, it was illegal in Britain knowingly to sell adulterated food as pure. The act also made it possible for local authorities to appoint food analysts. It has been heralded as the "first food law framed in the interests of the purchaser"—the first food law in which the rule of *caveat emptor* was replaced by *caveat venditor*.

But unfortunately, even its supporters acknowledged that in practice this new law was almost completely useless. The main trouble was that, while the act *permitted* local authorities to check food, it did not *require* them to do so. And since they were not required to do so, only two local authorities in the whole of Britain bothered to enforce the new law with any vigour. Another trouble was that the new law—following Hassall's lead here—defined adulteration very narrowly in terms of *intent*. This offered a great get-out clause for sellers, meaning that the dishonest had little to fear from the new legislation. Swindlers could be convicted only if it could be proven that they personally had an intention to deceive, which was, in most cases, virtually impossible to show; the grocer who sold fake tea could say that it had come to him in this form from China. And even if convicted, the most shopkeepers had to fear was usually a fine.[101] This new law effectively did nothing to protect the individual consumer. "No good resulted from it," wrote the distinguished food analyst Henry Letheby ten years later, "and it really stands upon the statute book as a dead letter."[102] It would take a lot more than this to stop the lies of the adulterators.

Mustard, Pure Food, and the Resilience of Commerce

In his evidence to the Parliamentary Committee, Hassall had dared to express the hope that government might take more responsibility for the food the public ate. The example he gave was mustard. It is fair to say that Arthur Hill Hassall was obsessed with pure mustard, or at least obsessed with the impossibility of obtaining it. Forty-two times, he had sallied out on to the streets of London to purchase some pure

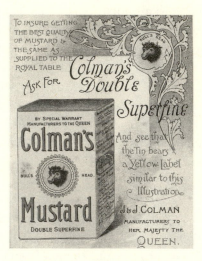

An early advertisement for Colman's mustard. Unlike the mustards of Hassall's day, Colman's was at last entirely pure.

mustard. Forty-two times he had been disappointed. On every occasion, he had been sold instead a mixture of mustard with "immense quantities of wheaten flour, highly coloured with turmeric."[103] He knew that some shopkeepers insisted that flour actually "improved" the mustard by lessening its pungency. But as Hassall remarked on another occasion, this was "sheer nonsense." "The volatile oil is the very essence of mustard, without it the article would be worthless as a condiment, and the addition of wheat flour does not in any proper sense neutralize it."[104] This universal adulteration of mustard was such a problem, he thought, that perhaps the best answer would

be "to compel the government to make its own mustard at a manu-
factory of its own"—government mustard. This way, universally pure
mustard could be assured.

The idea is reminiscent of a medieval guild monopoly, or of the
early socialists, who envisaged cooperative industry replacing that of
the market. But in 1855, Hassall's suggestion was so far beyond the
pale, it was laughable. Far from being about to embark on manufacturing its own mustard, the British government needed a lot of
pressure even to pass the limited and ineffectual anti-adulteration
law finally adopted in 1860. When Scholefield first brought a more
extensive anti-adulteration bill before Parliament, he was strongly
opposed by aggrieved grocers and coffee dealers—the so-called
shopocracy—who accused him of using "inquisitorial" methods
on his committee. The Commons, many of whose members were
themselves involved in trade, did not find any time for the bill in
the whole of 1858.[105] Later that year, a mass poisoning occurred in
Bradford, when more than two hundred people were laid low and
twenty killed by some lozenges. The lozenge-maker had intended
to adulterate his lozenges with plaster of Paris but had bought arsenic by mistake. It was only because of this Bradford lozenge scandal
that the government decided that the time had come to act. In 1859,
Scholefield introduced a weaker version of his original bill, and this
time it became law.

It is significant that the lozenge scandal that eventually impelled
the government to act involved actual poisoning rather than mere
deception. Everyone in British society could now agree that poisonous—or "injurious"—adulterations were a bad thing: no one really
wanted to drink foul water or to kill their children with coppered
sweets. More controversial was Hassall's view that *all* deceptions in
matters of food were bad, whether they were poisonous or not. His—
extremely stringent—definition of adulteration was "the intentional
addition to an article, for purposes of gain or deception, of any substance or substances, the presence of which is not acknowledged in
the name under which the article is sold."[106]

Not everyone agreed with this. A significant strand of opinion
still reckoned that a bit of harmless adulteration was defensible as

PUNCH, OR THE LONDON CHARIVARI.—November 20, 1858.

THE GREAT LOZENGE-MAKER.

A Hint to Paterfamilias.

"The Great Lozenge Maker," a *Punch* cartoon from 1858 referring to the poison lozenges scandal that year, which killed twenty people and led the British Parliament finally to adopt anti-adulteration legislation in 1860.

good for trade. At the Parliamentary Committee of 1855, a solicitor and chairman of a local board of health named Richard Archer Wallington gave evidence to this effect: "There is no understanding between the public and the seller that the seller shall give you what you ask for; *neither do I think it beneficial that it should be so*."[107] It was not a bad thing, Wallington thought, if people sold something as "coffee" that was really 75 percent chicory. "What the eye never sees, the heart does not grieve over." In Wallington's view, the only kind of adulteration that required definition was the poisonous kind. As for the other kind, it was "too extensive and undefinable" to deal with. It was best for commerce, thought Wallington, to ignore it. His position was echoed in a poem by the Irish man of letters William Allingham:

> *Adulteration is a form of Competition*
> Saith a British Manufacturer and Statesman;
> Shall we write upon his monument the sentence?
> There are many who will frankly hold it wisdom;
> There are some who interpret it more subtly,
> *British trade is more or less a form of cheating.*[108]

Cheating was not always the most profitable form of competition, however. Adulteration could be bad for trade, and honesty could be good. More dynamic branches of British commerce had spotted that the new public awareness of swindling could be turned to commercial advantage. After the 1860 act, the public were invited to submit suspicious food for analysis. The analyst Henry Letheby noted that in the nine years from 1860 to 1869, only fifty-seven articles were submitted, and of these, only twenty-six turned out to be of bad quality or adulterated. More than half the samples were "genuine articles" that "were brought to me with the knowledge of the dealers and with the evident intention of obtaining a certificate for trade purposes."[109]

Similarly, the upmarket grocer Crosse & Blackwell did an ingenious job of transforming Hassall's bad publicity into good publicity. In the original reports in the *Lancet*, a number of Crosse & Blackwell products were revealed to be thoroughly noxious. Crosse & Black-well's bottled gooseberries, for example, contained "a very consider-able quantity of COPPER. An iron rod immersed in the fluid in which the fruit is preserved became covered with *a thick coating of cop-per.*"[110] Copper was also present in Crosse & Blackwell's pickled gher-kins; while Crosse & Blackwell's anchovies were tinted with a toxic red colouring. As publicity goes, this was ruinous. When you are sell-ing "fine" goods to fancy customers, you don't want them to imagine their stomach lining clogging up with copper like an iron rod.

Thomas Blackwell went on the counterattack with a speed and bravura that were impressive even by the industrious standards of the day. Step one was contrite admission; Blackwell fulsomely con-fessed that, yes, he had indeed been guilty of using copper in his pickles and preserves in the past. Step two was to remove all copper and other toxic colours from Crosse & Blackwell products. The third step was to market this new purity of Crosse & Blackwell's pickles and preserves like crazy, as if it were a special improvement dreamed up by themselves, rather than something forced on them by bad publicity. Confession, contrition, remedy, ruthless self-promotion: as a marketing model for ensuring that all publicity can become good publicity, this has hardly been modified since.

At the 1855 Parliamentary Committee, Hassall indeed praised Crosse & Blackwell for having ceased to use copper as a greening agent. "For this change on their part, great credit is due to them, be-cause in many instances by doing so they run counter to the wishes and tastes of their customers." Hassall did have to admit, though, that the effect of this supposedly brave move had been "increased cus-tom."[111] Thomas Blackwell himself also spoke before the committee, betraying just how much he had benefited from his "honesty." Despite an initial falling off in sales, his customers were now "satisfied" that he had stopped using green colouring. "It is more to our interest to sell a pure article than an impure one, if parties will really take it," he admitted.[112] He used the example of anchovies. In Blackwell's own opinion, the undyed anchovy was "a very unsightly colour, an un-pleasant brown," and he knew that "many parties" would still prefer the red dyed kind (for some reason, anchovies were traditionally dyed brick red, often using lead oxide). "But the less attractive form of the article will be a sort of guarantee to the public of its being genuine." More discerning shoppers would learn to seek out Crosse & Blackwell products for their unattractive—hence natural—colours; the tan brown of their anchovies; the khaki green of their pickles; the muddy hue of their greengages. These muted hues would trumpet what every shopkeeper wanted every customer to believe: that their products were the most genuine of all. In the hands of Crosse & Blackwell, purity became a marketing device; and it has been so ever since.

From Adulteration to Packaged Purity

The 1860s and 1870s saw improvements for the British population, both as citizens and as consumers. Working men were given the vote and, with it, political responsibility. At the same time, the undesirable responsibility of the consumer for the quality of the food he or she bought was slowly shifted to sellers. Increasingly over this period, the interests of commerce converged with those of the consumer and of government to demand more honest food. William Scholefield MP died in 1867, the year of the Second Reform Act, but his activism was continued by Philip Muntz, another Birmingham MP and a fellow Radical. From 1868 onwards, Muntz, supported by Postgate and Letheby, doggedly brought new anti-adulteration bills before every session of Parliament. Initially, the government worried about losing the votes of shopkeepers. But by 1870, there was a new worry, which benefited Muntz's cause—the fear that Britain's reputation for adulterated food could damage its commercial standing in Europe. The foreign secretary in 1870 sent a questionnaire to all British consulates on the subject of adulteration, trying to gauge international opinion.[113] It was one thing to be known as a nation of shopkeepers, and another to be known as a nation of swindlers; this was bad for exports. That same year, the government reversed its position and committed itself to supporting Muntz's bill. Meanwhile, in 1871, a pressure group called the Anti-Adulteration Association (AAA) was founded to "amend and enforce the law against adulteration."[114] The association campaigned for the appointment of public

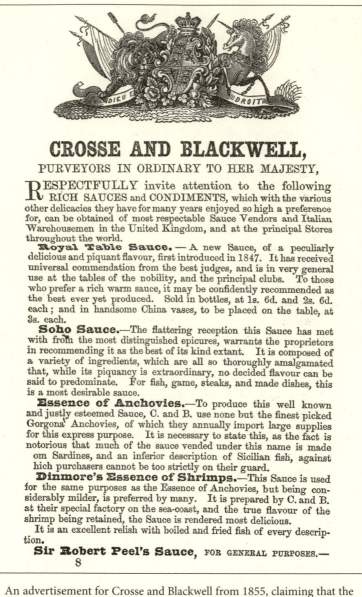

CROSSE AND BLACKWELL,

PURVEYORS IN ORDINARY TO HER MAJESTY,

RESPECTFULLY invite attention to the following RICH SAUCES and CONDIMENTS, which with the various other delicacies they have for many years enjoyed so high a preference for, can be obtained of most respectable Sauce Vendors and Italian Warehousemen in the United Kingdom, and at the principal Stores throughout the world.

Royal Table Sauce. — A new Sauce, of a peculiarly delicious and piquant flavour, first introduced in 1847. It has received universal commendation from the best judges, and is in very general use at the tables of the nobility, and the principal clubs. To those who prefer a rich warm sauce, it may be confidently recommended as the best ever yet produced. Sold in bottles, at 1s. 6d. and 2s. 6d. each; and in handsome China vases, to be placed on the table, at 3s. each.

Soho Sauce.—The flattering reception this Sauce has met with from the most distinguished epicures, warrants the proprietors in recommending it as the best of its kind extant. It is composed of a variety of ingredients, which are all so thoroughly amalgamated that, while its piquancy is extraordinary, no decided flavour can be said to predominate. For fish, game, steaks, and made dishes, this is a most desirable sauce.

Essence of Anchovies.—To produce this well known and justly esteemed Sauce, C. and B. use none but the finest picked Gorgona Anchovies, of which they annually import large supplies for this express purpose. It is necessary to state this, as the fact is notorious that much of the sauce vended under this name is made om Sardines, and an inferior description of Sicilian fish, against hich purchasers cannot be too strictly on their guard.

Dinmore's Essence of Shrimps.—This Sauce is used for the same purposes as the Essence of Anchovies, but being considerably milder, is preferred by many. It is prepared by C. and B. at their special factory on the sea-coast, and the true flavour of the shrimp being retained, the Sauce is rendered most delicious. It is an excellent relish with boiled and fried fish of every description.

Sir Robert Peel's Sauce, FOR GENERAL PURPOSES.—
8

An advertisement for Crosse and Blackwell from 1855, claiming that the "superiority" of its pickles is too well-known "to need any remark," even though just a few years earlier they were coloured green with toxic copper.

CROSSE AND BLACKWELL (*continued*).

Each bottle is labelled with a *fac-simile* of the Right Honourabl Baronet's letter of approval.

Strasbourg Potted Meats.—This delicious preparation far surpasses every description of Potted Meat yet introduced to public notice ; the flavour is full and rich, at the same time being so mild and bland, that the most delicate and fastidious palate is pleased. It is also very easy of digestion, and adapted to weak stomachs.

Calves' Feet Jellies.—Consisting of Orange, Lemon, Noyau, Punch, Madeira, and Calf's Foot. These are sold in convenient sized bottles, and their use is attended with a great saving of trouble and inconvenience ; besides which, they ensure the certainty of the Jelly always being of uniform excellence and flavour.

They are now in almost daily consumption in many families, and are very highly approved.

Pickles OF ALL DESCRIPTIONS.—The superiority of these is too well known and appreciated to need any remark.

Fruits IN BOTTLES.—Preserved *pure* for Tarts ; available when fresh fruits are not in season.

Syrups OF VARIOUS KINDS, FOR FLAVOURING ICES ; or which, if diluted with iced or spring water, produce a cool and refreshing beverage.

C. and B. are also agents for the following, made by Monsieur SOYER :—

'Soyer's Aromatic Mustard.'—A most exquisite combination of the genuine Mustard seed with various aromatic substances: infinitely superior to all other preparations of Mustard.

Soyer's New Sauces.—One of a mild description for the ladies, and another of the same flavour, but warmer, for gentlemen.

Soyer's Relish.—With reference to this Sauce, the *Observer* remarks :—

'M. Soyer is a culinary artist as profound as he is versatile ; nothing comes amiss to him. No foreign *cuisinier* ever tickled the Saxon palate so successfully. He is a great man ; and the ill-cooked mutton chops that lost Napoleon the battle of Leipsic, would have produced a very different effect if Soyer had dished them up in his Magic Stove, and rendered them thoroughly light and digestible by his appetizing Relish.'

C. and B. consider it important to state, that the whole of their manufactures are prepared with the most scrupulous attention to cleanliness and purity. The utmost precaution is taken in every instance to prevent contact with copper, or any other pernicious metal ; and to ensure this end, they have at a great expense fitted their factory at Soho Square with a number of Earthenware Steam-pans, and in addition have had a large Silver pan made, in which to prepar the most delicate of their productions.

WHOLESALE WAREHOUSE—

21, SOHO SQUARE, LONDON.

9

analysts who could put the law against adulteration into practice. In 1872, this happened, when Muntz's Adulteration of Food and Drugs Act was passed.

This act was a vast improvement on that of 1860. It was now an offence to sell a mixture containing ingredients designed to add weight or bulk, unless explicitly stated as such. You could still sell chicoried coffee or "cassia'd" cinnamon, but only if you said that was what you were doing. New local officers—public analysts—were given extensive powers to procure samples for analysis. It was clear that "the new Act was a great advance on its predecessor, more extensive in its scope and more vigorous in its enforcement."[115]

There were still some teething troubles. Grocers protested vigorously at the fact that the new law no longer made knowledge of adulteration necessary to secure a conviction. An honest shopkeeper with dishonest suppliers could thus be punished for a crime he had no knowledge of. There was now a genuine law of *caveat venditor*. After the Conservatives won the general election of 1874, they appointed a select committee to examine the new law; this committee concluded that "some respectable tradesmen" had indeed been given undeserved penalties. There were doubts, too, as to the competence of some of the new public analysts, most of whom lacked Hassall's skill with a microscope and could not agree among themselves about what exactly constituted adulteration: what should the minimum percentage of fat in milk be? Did the colouring of tea count as adulteration? These problems were addressed in the Sale of Food and Drugs Act in 1875, which dealt with some of the grocers' grievances as well as giving a clearer understanding of what does and does not constitute adulteration. This act still forms the basis of British food law.

By the 1880s the worst horrors of the demon grocers were past. The improvement in British food during this decade has been called "spectacular"; "by the end of Victoria's reign the consumer generally received his bread and flour, his tea and sugar, as pure as he could wish."[116] In 1872, thirty-six out of forty-one samples of tea were found to be adulterated with additives such as Prussian blue, substitute leaves, China clay, and sand. By the late 1880s, adulterated tea was the exception, rather than the rule. Similarly, the work of the public analysts showed that the percentage of adulterated bread had dropped from 7.4 percent in 1877 (already much lower than the 1850s) to 0.6 percent in 1888.

British food was safer now than at any previous time in the nineteenth century. Yet opinion often lags far behind fact, and the British public continued to believe that their tea and bread and pickles were regularly tampered with—as G. K. Chesterton's 1914 "Song against Grocers" shows—long after they really were. This mood of nervousness was a gift for manufacturers and advertisers, who created new branded goods in this era, each one promising a unique trustworthiness. Lyle's golden syrup, Bovril beef extract, Horlicks malted milk, Bournville chocolate; all these were first marketed in the 1880s or 1890s. In the 1840s, advertisers had pushed the ideal of the "genuine"; now, no less spuriously, they pushed the ideal of extreme purity, with the implication that whatever was concealed in the packet was safer than the food lying open to view in the marketplace. Here lay the beginning of the packaging revolution.

Thus by a strange twist, the war on swindling had given new opportunities to the manufacturers of processed foods. There were tempting profits to be made, as we may see by returning one last time to the contradictory figure of Arthur Hill Hassall. Despite his invectives against advertisers, Hassall attempted on at least three occasions to market "pure" foods himself. After suffering from a severe lung infection, he had spent the period from 1868 to 1877 establishing a special hospital for treating patients with pulmonary conditions on the Isle of Wight—the Ventnor Hospital. On the side, he continued to analyse adulterated foods and wrote frequent letters to the press on the subject, pointing out, for example, the hazards of lead colouring used in postage stamps.[117] It was noble work, but not profitable, and it is clear from his autobiography that he was desperate for wealth. Hence, his ventures into the world of commerce, the very world which he had criticized so effectively in the past.

Hassall believed, however, that what he was doing was different. When selling *his* foods, he would "always circumscribe the statements advanced" on the food's behalf "within the limits of sober

147

BIRD'S CUSTARD POWDER *makes a perfect high-class Custard at a minimum of cost and trouble. Used by all the leading Diplomées of the South Kensington School of Cookery. Invaluable also for a variety of Sweet Dishes, recipes for which accompany each packet.*

NO EGGS!   NO RISK!   NO TROUBLE!

An advertisement for Bird's Custard, which openly contrasts the reliability of packaged food with the risks and bother of making a real egg custard.

truth." His first attempt at selling a pure food was in the 1860s—an unpalatable-sounding product called "flour of meat." The idea was to create "an article analogous to wheat flour in fineness" but made from real meat, bought from Spitalfields, minced, pulped, dried, and pulverized. Hassall considered this product a "real triumph of manufacture" and marketed countless varieties: Flour of Meat Mixture for Soups, Flour of Meat Food for Children, the Aged and Invalids, Flour of Meat Cocoa, Flour of Meat Biscuits, Flour of Meat Lozenges. At first, Hassall said, "orders came in freely" and "success seemed assured." But then there were complaints that the flour did not keep well, and Hassall had to admit that "these complaints were to some extent well founded."[118] That was the end of that. The company folded. Next, in 1875, he launched a food for invalids made from baked wheat flour, malt flour, diastase, and cerealine, but this—despite its very "pleasant" flavour, according to Hassall—was no more successful.[119]

His final and most ambitious food project was set up in 1881. By this time, he observed, gross adulteration was far less common, and people therefore wanted more than food that was merely not adulterated. They wanted "to secure the absolute purity of our food, especially of that required by invalids."[120] What if a company could guarantee the purity of all the articles it manufactured? Wouldn't this prove to be a "great success" (and of course a source of profit to Hassall himself)? Hassall persuaded a gentleman he knew to provide the start-up costs, and "The Pure Food Company" established its headquarters at 4 Princes Street, London, Hassall being assisted by a friend, Otto Hehner, who worked free of charge. All the ingredients used were "of the best quality." The water was "softened and purified." All the meat was bought daily from Smithfield. The company would focus on making "solidified or concentrated beef tea, the same with the albumen of the meat, the same with arrowroot, pure beef jelly or essence, albumenous or fibrinous meat lozenges, milk food for infants, the same for children and invalids, pulsella [presumably some kind of lentil or pulse], cooked and in part pre-digested; extract of coffee, prepared by a special process and preserving the full aroma of the berry." Finally, there would be extract of coffee and chicory,

149

but scrupulously labelled as such, so that no coffee lover would be deceived. Hassall and Hehner threw themselves into the project. The company made a prospectus of its products "stating truthfully the facts without exaggeration." Yet the sales were never sufficient to cover costs and, as Hassall put it, "this well intentioned venture came to an end." Hassall headed off to a frugal retirement in San Remo, where the climate agreed with his poor nervous constitution.

Never one to take a disappointment well, Hassall knew exactly what conclusions to draw from the failure of the Pure Food Company, blaming the ingratitude of his public and the hopelessness of competing with commercial puffery: "So much for the discernment and appreciation of the public, who are ever crying out for, and writing about, the purity and quality of food. Had the prospectus contained extravagant and exaggerated statements and assertions and had a fortune been spent in advertisements, then the result might have been different."[121] It is possible, though, to take a different moral from Hassall's ill-fated Pure Food Company. "Fibrinous meat lozenges," "flour of meat." "pre-digested pulsella"; these may all have been purer than pure, but none of them are recognizable *foods*. In his frantic search for purity, Hassall had lost sight of the reason why adulteration mattered in the first place; unlike Accum, with his instinctive zest for life, Hassall became so obsessed with deception that he had forgotten the need for good food.

The case of Hassall shows how easily the fight against swindling can turn into a futile quest for purity, which ultimately does little to aid the cause of the individual consumer. History shows that adulteration flourishes most when people no longer trust their senses; when they lack a firsthand knowledge of what is good. Science ought to be a complement to this intimate knowledge, not a substitute for it. Hassall succeeded in achieving the shift from "buyer beware" to "seller beware." But the latter code has its dangers too. If the seller assumes responsibility for the integrity of food, it can mean that buyers no longer feel they need to keep their eyes open; that they can trust in the superior knowledge of food technicians. Hassall's Pure Food Company was asking people to forget their own senses and put their faith only in the science of his sanitized products—the same siren call that industrial packaged goods have offered to consumers ever since, the promise of clean food whose purity is guaranteed by the brand. This story was to be played out most dramatically in the United States, where the quest for purity and the drive of the free market were to do battle, and occasionally to join forces, well into the twentieth century.

4

PINK MARGARINE AND PURE KETCHUP

Mary had a little lamb,
And when she saw it sicken,
She shipped it off to Packingtown,
And now it's labelled chicken.   —*New York Evening Post* (1906)

To be cheated, fooled, bamboozled, cajoled, deceived, pettifogged,
demagogued, hypnotized, manicured and chiropodized are privi-
leges dear to us all . . . Americans like to be humbugged.
                          —Harvey Washington Wiley (1894)

At the time when Britons were waxing most desperate about the falsification of their food and drink, in the 1850s, they would some-times look more than a little enviously across the Atlantic at condi-tions in America, where many people still raised most of what they ate and had little need of protection from the nefarious ways of the adulterators.[1] Americans too could reflect with some complacency that for all the disdain with which their manners, politics, and cul-ture were treated by old Europeans, their food production was the envy of the world. Yet in the decades after the Civil War (1861–65), all that changed.

Rapidly, American food suffered the same deterioration in quality that British food had suffered in the early nineteenth century, and the reasons in each case were very much the same. America was finally making the shift from a predominantly agricultural society to an industrial nation. With big industry came new technologies—new ways of tampering with food, and new and powerful markets for selling it in. In the 1870s, big manufacturers began employing industrial chemists to invent new aids to deception—deodorants, dyes, flavours, crispers for flabby foods, and softeners for hard foods—which left consumers thoroughly confused about what they were eating.[2] By the 1880s, "the whole system of food supply was assuming a different appearance," with an increasingly urbanized population calling for cheap new processed foods.[3] Things had got so bad that by 1892 a U.S. senator, Algernon S. Paddock, could complain that "the devil has got hold of the food supply of this country."[4]

This epidemic of adulteration, occurring later, played out differently from the British experience a generation earlier. Everything was on a grander scale, and the battles between the swindlers and the purifiers were bigger too. Whereas the English fight had been one of science against science—and often seemed like gentlemen against gentlemen—in America, it was a fight of commerce against commerce. The Big Boys of the Beef Trust and the whiskey rectifiers were keen to label anti-adulteration campaigners as cranks and social misfits. But other, shrewder business minds, such as the grocer James Thurber and the manufacturer Henry J. Heinz, spotted the benefits of allying themselves with the cause of pure food. They discovered that there was serious money to be made out of cleaning up their act, yet the suspicion often remained that it was still just an act, and that the new "pure food" labels were just as misleading as the old dyes and pomades.

The nature of the political debate on swindling was different, too. In Britain, the central political question had been how far the government should intervene in the market. But in the United States, this was complicated by the fact that there was more than one kind of government—state government and federal government. It was impossible to discuss whether there should be better laws to protect the nation's food without coming up against the old chestnut of states' rights versus federal power, the perennial squabble that had affected every political question in America since the Constitution was ratified in 1787.

On the antifederalist side were those like Congressman Adams of New York, who commented in 1884 that food regulation was "a local matter belonging to the states"—which meant that some states were ahead of their time but many others lagged behind.[5] On the federalist side were those like Harvey Washington Wiley, the pure-food crusader, who insisted that the quality of food eaten by the average consumer could not be guaranteed unless there were federal laws to regulate the trade in food between states. The eventual triumph of the federalist point of view—in the Pure Food and Drugs Act of 1906—showed how crucial food was in fashioning modern America, one nation under God, commerce, and federal government.

Another difference was the tone in which food swindling was discussed. The mood of the *Lancet* and of Hassall was sceptical, understated, reasonable, sometimes infuriatingly so. By contrast, discussions of food in the United States were often infused with messianic zeal, much of it derived from the Temperance movement. Eating and drinking the wrong things became bound up with the idea of sin. Whereas Hassall worried about the purity of the mustard and coffee freely on sale in London, for many of the American anti-adulteration campaigners coffee and mustard were ipso facto impure, evil stimulants, even if sold in unadulterated form. They didn't just want their food to be reliable; they wanted it to be pristine.

Milk and Alcohol

The Pure Food movement proper did not really take off in the United State until the 1870s; but it was preceded by two decades of periodic food scares in which the newspaper-reading public of the big cities was alarmed with reports of "vile mixtures" and frauds. By far the most prominent of these scandals was the swill milk affair of the 1850s, which began in New York City. Yet New York was seen by many people, including many of its own inhabitants, as a case apart. The result was that public horror was not matched by meaningful political action. As with many public health scandals, the first instinct among the politicians was to try to get everything back in proportion—simply to calm people down—so that decisive action was a long time coming. In the meantime, some of the old faith in the purity of the American way of life began to erode. If you couldn't trust milk, what could you trust?

Milk has always played a troubling role in the American psyche: it is an American staple, but one that is capable of provoking deep unease.[6] The mainstream image of milk remains very positive to this day, linked to the childhood innocence of "milk and cookies," the belief in milk as a kind of white nectar, wholesome, calcium-rich, and pure (an image heavily promoted by the National Dairy Council). Yet this has been offset by periodic bouts of anxiety about whether milk is all it seems. There have been frequent backlashes against the apparent inviolability of milk's place in the American diet, with sceptics arguing that milk is neither the best nor the only source of calcium, and that milk consumption is linked to a number of serious diseases. Sometimes, the tone is distinctly alarmist, if not downright apocalyptic. *Milk: The Deadly Poison* was the title of Robert Cohen's antimilk tract of 1997.

There was a comparable divide in 1850s' New York. Milk retained a reputation as potentially the most perfect of foods, but one that had sunk in the wicked den of the city to a thoroughly debased state. In truth, during the period after the birth of the modern city but before the advent of pasteurization and the refrigerator, milk could be a lethal substance. In his *Essay On Milk* from 1842, Robert Hartley had warned that milk's purity could be catastrophically tainted if the milk were produced under the wrong conditions, yet these taints could prove very hard to spot. The warning proved to be eerily prophetic.

In the first half of the nineteenth century, when urban populations were not so dense, city milk was generally supplied by cows pastured on grass in the city itself. But as houses crowded in, pasture space shrank, and New York had to find new ways of procuring milk. From 1842 onwards, the year of Hartley's essay, some fresh milk was brought to the city each day via rail, from rural areas such as Orange County. Most of the city's milk, however, was so-called swill milk or slop milk—milk from cows kept in dairies attached to breweries and distilleries. These cows were kept in vast, darkened cowsheds and fed the hot grain mash left over from distilling. By 1854, it was estimated that there were thirteen thousand cows fed on swill "and living a life of appalling misery," while producing a milk that was blamed for the deaths of thousands of children every year.[7] Slop milk was thinner and more watery than country milk—its fat content was too low for it to be made into butter or cheese—yet by 1850 the majority of New York milk was produced in this way. It also bore the distinct taint of alcohol. In a society where temperance meetings were becoming common, swill milk started to attract strong moral disapproval. "With swill milk for children and swill liquor for men, no wonder that we are so healthy and hearty a race," wrote one sarcastic commentator, appalled by the tide of physical and moral turpitude he foresaw sweeping the nation.[8] But others saw things in even starker terms: swill milk wasn't debauched; it was deadly.

"Death in the Jug" was the headline in a *New York Times* report on swill milk from 1853, borrowing Accum's phrase. A campaigner named John Mullaly had just brought out a pamphlet, *The Milk Trade in New York and Vicinity*, anatomizing the horrors and swindles of the city's milk, pointing out that much of the milk sold as "country milk" was swill milk in disguise. The *New York Times* writer linked the state of milk—"abuses that poison the very springs, of life"—with the poisonous quality of city life in general. This was an era of filth and chaos. The opening of the Erie Canal in 1825 had made New York the economic heart of the nation, but the city had yet to develop a civic politics to regulate the new economy. There were distinct echoes of Accum's London of 1820—in both cities industry had galloped forward without either the social responsibility or the political institutions to rein in the mess. "Our streets are never cleaned. Our police force is inefficient, and our thoroughfares are obstructed by a thousand disagreeable obstacles. We walk in mire, sleep in fear and run the risk of being trampled to death by infuriated oxen in our principal highways."[9] And to top it all, the substance sold in Brooklyn and Manhattan as "pure milk" was anything but. The *New York Times* attributed the deaths of up to eight thousand children a year to this foul fluid.[10]

This was an overstatement; but it is true that there was a direct connection between high infant mortality and the disgusting conditions

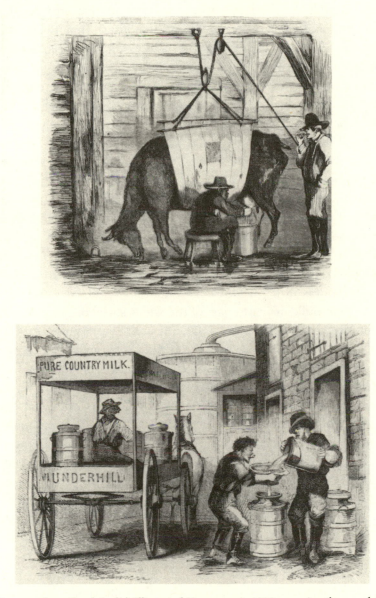

Cartoons from *Frank Leslie's Illustrated Newspaper* in 1858, exposing the scandal of swill milk in New York City.

in which milk was produced. It was unfortunate that breastfeeding of children under one year of age declined at just the point in history when cow's milk was least safe to drink. The economic need to work soon after confinement compelled many women to give up nursing their own babies. But where in the past they might have used a wet nurse, increasingly they turned to substitute milks. The practice of wet-nursing had begun to decline by the start of the nineteenth century.[11] By the 1860s, "artificial feeding" of babies was on the rise, boosted by the invention of feeding teats made from India rubber, and promoted by many physicians as the modern, hygienic thing to do. In 1869, the German chemist Justus von Liebig (1803–73) marketed a soluble breast milk substitute, made from wheat flour, dried cow's milk, malt, and some bicarbonate of potash to make it less acidic.[12] Those who could not afford it often fed their babies on cow's milk, sometimes mixed with water and sugar. When the milk in question was swill milk, the consequences could be devastating.

The swill milk affair was part of the wider social disgrace of high infant mortality. From 1870 to 1900, one out of every three deaths in the United States was of a child under five. Between 38 and 51 percent of infant deaths were from infectious diseases, and of these, half were diarrhoeal infections, which were especially connected with the consumption of bad milk.[13] Death from diarrhoea peaked in July and August when bacteria in the—often already filthy—milk multiplied more rapidly. Many theories have been suggested for the high rate of infant death: poverty, lack of education, overcrowding, bad sewerage. In 1909, one doctor suggested that "dirty pacifiers, overdressing or pickles" might be to blame.[14] It was bad milk, however, that was the critical factor: it was not poverty per se that caused babies to die, so much as the fact that poor mothers were more likely to be duped into buying cheap, poor-quality milk. When the infant mortality rate finally dropped, in the first two decades of the twentieth century, it was in direct response to the fact that the milk supply had at last been cleaned up.

Not only in America was milk a problem at this time. Throughout western Europe, city milk was a dangerous substance. In Paris, during the siege of 1870, infant mortality fell by 40 percent, because mothers were forced to suckle their children instead of relying on cow's milk.[15] London milk was no better. In *David Copperfield*, Dickens has the character of Mr. Littimer complain about the cocoa he is given in prison. "If I might take the liberty of saying so, sir, I don't think the milk which is boiled with it is quite genuine; but I am aware, sir, that there is a great adulteration of milk in London and that the article in a pure state is difficult to be obtained."[16]

As much as half or even three quarters of London milk was adulterated in the mid-nineteenth century. First, it was watered down (often with contaminated water). Then, to compensate for the bluish colour, it was thickened with flour, sweetened with carrot juice, and coloured with yellow dye. Trade dyes were sold under names such as "Silver Churn" or "Cowslip Colouring."[17] It was often dosed with chemicals, with such names as "Preservitas" and "Arcticanus," especially in the summer months, to stop it souring so quickly. These chemicals were especially dangerous, because rather than stopping the decomposition of the milk, they merely masked it, fooling you into thinking that you were drinking something reasonably fresh. "Four day old milk with an overdose of chemicals was hardly the ideal infant food," comments one historian. No wonder some consumers asked for the cow to be brought to the doorstep, and milked in front of them, something that still happens in some South Asian cities.[18] But even the horrors of Victorian London paled next to the scandal of New York swill milk.

The *New York Times* observed that most housekeepers, if asked where they bought their milk, would say they got it from "the regular milkman, who imports it direct from Orange County, Westchester or Connecticut," as the case may be. These "poor deluded" housekeepers might have "vague visions of delicious pastoral dairies, where the fresh perfume of the country fills the nostril with a pleasant sense— of milking hour in the meadows, when rosy-cheeked girls go out with their pails, and are welcomed by the waiting cows, who know every face among them."[19] "Dreams all!" The reality was that a large proportion of what was sold as "Orange County" milk was heavily watered down, and much was not country milk at all. Mullaly had discovered that 90,000 or so quarts of milk entered the city each day, yet by some "magical process" this had increased to 120,000 quarts at the time of delivery, which confirmed that many milkmen diluted their milk by at least a quarter with other liquids. Yet bad as this was, the *New York Times* argued that it would have been better for mothers and children if the worst they had to fear was country milk "adulterated with chalk, water and a little molasses." Such milk was harmless compared with swill milk—of which up to 160,000 quarts were produced daily[20]—filled as it was with "contagion" and "reeking with animal poisons." Oh for a little honest swindling compared to the horror of what Mullaly saw when he went to visit the swill dairies.

Mullaly had been to West Sixteenth Street, between Tenth Avenue and the North River, to visit the premises of a rich distiller called Johnson, "who 'boards' two thousand cows at six cents a day each and clears forty thousand dollars a year by his nefarious business." The "stench" from this place could sometimes be smelt a mile off, ruining the quality of the life of those living in the vicinity. The air in the stables was so foul that when health wardens went to inspect them in 1854, as a result of Mullaly's pamphlet, they were forced "to suspend the inspection for a time to recover from its sickening effect upon them."[21] The cows were crammed in, six or seven hundred to a single filthy stable, and forced to stand more or less constantly over the swill trough, except when lying down on the dung-encrusted floor. Mullaly reported:

> When the swill is first served it is often scalding hot, and a new cow requires some days before it can drink it in that condition. It instinctively shrinks from the trough when the disgusting liquid is poured in, but in the course of a week or two, it becomes accustomed to it and finally drinks it with evident relish. The appearance of the animal after a few weeks feeding upon this stuff is most disgusting; the mouth and nostrils are all besmeared, the eyes assume a leaden expression, indicative of that stupidity which is generally the consequence of intemperance.[22]

Mullaly observed how little care these poor drunken beasts were given by the humans who looked after them; and by implication how little care was given to making the milk they produced wholesome. The milkers did not bother washing their hands but would use their unclean fingers to pick out bits of conspicuous dirt from the milk. The cows' udders were frequently ulcerated, but they would be milked regardless, "and the milk mixed in with the rest." They would continue to be milked up to the hour of their death, even when they could barely stand. After it left Johnson's premises, the milk would often be tampered with still further by small dealers, as the *New York Times* described:

> To every quart of milk a pint of water is added; then a quantity of chalk, or Plaster of Paris to take away the bluish appearance produced by dilution. Magnesia, flour and starch to give it consistency, after which a small quantity of molasses is poured in to produce the rich yellow colour which good milk generally possesses. It is now fit for nurseries, tea-tables, and ice cream saloons, and is distributed over the City, insidious, fatal, and revolting poison.[23]

The city police did remarkably little. As the *New York Times* noted, if "any of us citizens" decided to keep a mad lion on their premises, the authorities would intervene, whereas Johnson was allowed to add to his income by "poisoning the population in two ways"—first, by distributing the "diseased milk" itself, and second, by generating a "horrible and fetid vapour."[24]

Yet for decades, swill milk was the scandal that would not go away. In 1858, Frank Leslie's *Illustrated Newspaper* undertook yet another exposé of the distillery dairies, which showed once again "the filthy condition of the cow-stables, the misery of their occupants and the brutality of their owners."[25] Readers were shocked to learn of the effects of distillery swill on the cows—rotting teeth, tails that fell off, ulcers. Once again, our friend Mr. Johnson of West Sixteenth Street was identified as one of the chief culprits. For the sake of appearing to do something, Tammany Hall sent Alderman Michael Tuomy to "investigate." Tuomy was a pugilistic butcher turned politician.[26] In lieu of embarking on a proper investigation, he chose to drink a few glasses of whisky with the dairy owners and then did everything he could to shield them, and to accuse those who disliked slop milk of simply being "prejudiced," announcing on the basis of his extensive investigations that swill milk was as good for children as ordinary

161

milk.[27] Swayed by Tuomy, a committee of the common council of New York concluded that distillery dairies were "as clean as it was possible for such places to be" and that it was impossible to cite a single instance of a child dying from drinking swill milk.[28]

Tuomy soon became caricatured in the *Illustrated News* as "Swill Milk" Tuomy; he responded in characteristic fashion by arresting the editor. Meanwhile, swill milk not only continued to be sold in New York, but spread to other cities such as San Francisco, Chicago, and Philadelphia.[29] The Annual Report of the New York City Inspector of 1860 showed that the production of swill milk continued "without the slightest abatement or any sign of diminution" and would continue to do so "unless the adoption of measures of the most stringent character is resorted to."[30] Finally, in 1862, New York did pass a law prohibiting the sale of adulterated milk and the keeping of distillery dairies, and requiring milk dealers to label cans and vehicles with their names; but neither the measures nor the will to enforce them were sufficient, and swill milk scandals were still common in the 1870s, as was the practice of watering milk.

What made milk such an especially tricky substance to regulate and such an easy substance to falsify was its unpredictable, natural qualities; it was both a breeding ground for disease and a highly variable product.[31] Bacteriology was a science that was only just emerging. Analysis of distillery milk showed it to be 89 percent water on average compared with 86 percent for Orange County milk; but levels of water and cream in milk fluctuated hugely. If consumers found their milk less creamy than usual, they could be left in doubt as to whether this thinness was natural, or a swindle. As a report of 1873 wryly commented:

Recent investigation has shown that the cow is in conspiracy with the milkman to water our milk. . . . Milk contains in its natural state, a very large percentage of water. When the proportion of cream varies from twenty-three to five per cent; when the same animal will yield in the morning milk having 11.5 per cent of cream, and in the afternoon milk containing only five per cent of that ingredient which people perhaps foolishly prize—who is to decide whether the cow or the milkman has contributed the extra quota of water?[32]

In order to obtain milk that was reliably safe to drink, the consumer had eventually to abandon the dream of milk in its natural, pastoral state, and trust instead to the sterile hand of science. It was only the invention of the portable milk bottle and systematic pasteurization, finally coupled with much better regulation, that would secure safe milk for Americans, even in the swill-city of New York. From 1893, a philanthropist named Nathan Straus (1848–1931) financed low-cost milk depots selling pasteurized milk to poor New York families.[33] As a result, infant mortality rates finally dropped.

But the swill milk scandals of the 1850s onwards planted a seed of doubt in the public mind about what they were eating and drinking, and soon concerns about swindling and bad food had spread far beyond the dairy. In the 1870s, newspaper articles began appearing that attacked the entire food supply as unsafe and iniquitous, similar to the scaremongering tracts in Britain of the 1840s, with the same litanies of fraud: chicory in coffee, lead in sweets, copper in tea. In 1871, one report compared the "infinitesimal" degree of adulteration in former years with its present ubiquity, attributing the rise to the prevailing money obsession of the times.[34] In 1872, another journalist wrote that

> Adulteration of food is an evil so widespread, so tempting to its perpetrator, so difficult of detection by its victim, that most people have come to accept it nowadays as practically inevitable. We cannot all be analytical chemists, and perhaps on the whole, so far as the pleasures of eating and drinking are concerned we may be thankful that we are not . . . if people only knew what abominable messes they are constantly putting down their throats under the most innocent disguises, the healthiest appetites might well revolt.[35]

There was vague speculation that "nine-tenths" of the food supply was not what it "pretends to be."[36] Horror stories multiplied—of sulphuric acid in "spurious vinegars," for example, leading to sore gums, "paralyzation of the nerves of taste, piles and other dire calamities" including "hiccoughs, 'ahems!' and other violent clearings of the throat."[37]

The effect of such panicky claims, endlessly repeated, was the same as it had been in Britain: readers got bored with the subject of adulteration before anything had been done to combat it. In 1881,

Dr. Charles Smart was asked by the National Board of Health to investigate the problem of adulteration. He reported that "sensational writing" had made people inclined to "discredit the whole thing, except perhaps that milk may sometimes be watered" (an idea that people had become used to).[38] Too many writers, he complained, had applied Hassall's work to the American situation without any thought for how the circumstances there might be different. Smart himself pointed out that, thanks to the widespread American habit of grinding coffee at home (a throwback to the times when grocer shops were few and far between), American coffee was much less adulterated than the British ground variety. American cornmeal and wheatmeal were also purer, he blithely assured the public.

Yet, based on the raw data of Smart's own analyses, many American foods were found to be as adulterated as the British, if not worse. Smart uncovered "cinnamon" that was a mixture of cassia, almond shells, corn, wheat, allspice, and beans; yellow coloured candies poisoned with lead; ground allspice—which was hardly adulterated in England—mixed with breadcrumbs, "woody tissues," and turmeric. Sugar, meanwhile, was widely adulterated with glucose, which had not been the case in England. Later analysis by other government agencies showed lard preserved with caustic lime and alum; cheese containing mercury salts; canned foods laced with copper, tin, and preservatives.[39] Many of these additions were motivated by greed; but the difficulty in some cases, thought Smart, was that "the dealer himself appears to have lost knowledge of the characters of the pure article."[40]

This confirmation of people's worst fears, coupled with a deep-seated reluctance on the part of the authorities to do anything about it, meant that the battle to restore knowledge of what was good and pure in food was often left in the hands of America's women's groups, though their idea of the "good" and the "pure" was perhaps loftier than many people could live up to. In the 1880s, in isolated pockets throughout the United States, many different women worked to teach other women how to avoid poisoning their families. Isabel Churchill, a clubwoman from Denver, Colorado, argued that "the quality of food stuff should be a question of paramount interest to every house wife, and therefore club women should all be concerned with pure food legislation."[41]

They met in church halls and lecture halls, in club houses and private homes. In New York in 1884, when manure rotted outside slaughter-houses from Forty-third to Forty-fourth street, creating an unbearable stench, it was a group of fifteen "ladies" who campaigned to clean it up.

That same year, further west, Ella Eaton Kellogg organized a meet-ing of the Michigan Women's Christian Temperance Union to discuss pure food. They met at Battle Creek, Michigan, where Ella's husband, James Harvey Kellogg, ran his famous sanatorium. Ella Eaton Kellogg proclaimed that women should be taught to eat unadulterated and uncontaminated food, rather than rushing off to buy quack cures. "The home is woman's citadel," she explained. "It is here disease most often threatens." Like many temperance women, Ella believed that abstinence from liquor was closely linked to abstinence from com-promised or artificial food. Her definition of adulterated food was extremely broad, including not just spurious concoctions but also refined sugar, and "stimulating" condiments such as pepper and mus-tard.[42] This view of mustard as a narcotic might seem a little surpris-ing; it needs to be seen in the context of American Temperance diets from Sylvester Graham (1794–1851) onwards. As early as the 1830s, Graham—father of the Graham cracker and of Graham bread—had set out a vision of food in which condiments, spices, sugar, salt, cof-fee, tea, and even shellfish were seen as damagingly "stimulating" (in contrast to the "Edenic foods" of fruit and vegetables eaten by Adam and Eve).[43] So while Hassall was worrying about the impossibility of obtaining pure mustard, the Grahamites were worrying that people were eating mustard at all. Later pure-food campaigners such as Ella Kellogg shared Graham's basic presuppositions but restated them at a time when the fight against impure foods seemed all the more urgent. By the 1880s, America had become a land of domesticated narcotics and nostrums, of "soothing syrups" for babies that turned out to be opium (there was a scandal in 1888 when Mrs. Winslow's "soothing syrup," which had been promoted in church magazines as "invaluable for children" to help them sleep, was exposed as opiate-based).[44] The women's movement for pure food was a reaction against all this.

Ella Kellogg's extreme ascetic position was also a reflection of her fierce moral absolutism and that of other campaigning women.

What gave their language such power was its vision of adulteration as sin. But this also led them to more sweeping assertions than Accum or Hassall ever made, as the language of taint became equated with the corruption of the entire body. In 1885, Ella Hoes Neville, a clubwoman, wrote that "the adulteration of food is a sinful dealing, worse than any short weight or dishonest fabric. Give us short measure and we lose; give us adulterated food and we die."[45] By the same token, in the 1890s the editor of *Club Woman* magazine insisted that every single swindler of food should be eliminated. Every seller of fake sugar or "renovated butter" poisoned the "avenues of trade." "As a mere drop of poison will pollute a whole quart of milk, so will one dishonest merchant or manufacturer corrupt the whole industry."[46]

Inexorably, the fight against adulteration became bound up with the language of sin and redemption. The vegetarian pure-food campaigner George Thorndike Angell, president of the Massachusetts Society for the Prevention of Cruelty to Animals, fought for a broad national law to protect the American people from bad food.[47] Angell saw his job as being equivalent to a "preacher in a pulpit." Now that slavery was abolished, good men had a duty to fight the slavery of cattle kept in inhuman conditions for food, and the mass poisoning of the populace by manufacturers who were little better than "the highwaymen who rob and murder."[48] Angell saw humanity sinking fast in the "unfathomable sea of adulteration," and he saw it as his job to rouse them to save themselves. For Angell it was no good just correcting this or that aspect of modern industrial food: it was all bad, all a giant conspiracy of evil. Industry responded to Angell in kind. There was no use taking him on point by point: everything that came out of his mouth was the ranting of a crank. Thus, each side in the adulteration debate encouraged the excesses of the other, leading to a stalemate within which there was no true communication.

Even among the people, both in business and in government, who broadly supported more regulation for the food trade, the hectoring of the radical fringe was deeply off-putting—exaggerated, sensational, and just plain irritating. Pure-food crusaders such as Angell or the temperance women were often accused of being "fanatics, socialists, cranks."[49] Nevertheless, as the nineteenth century neared its end, a number of individual states were eventually prompted to pass new laws in an attempt to safeguard food quality. In 1874, Illinois had passed one of the first state laws to apply to pure food in general rather than individual foodstuffs; New York produced a general pure food and drugs act in 1881; Massachusetts followed in 1882, largely thanks to the efforts of Angell. What failed to materialize, however, was any national effort to address the problem. What was needed was a good, old-fashioned conflict of interest that drew in the U.S. government. That came with the arrival in North America of an exciting (or deeply sinister, depending on your perspective) new product from the laboratories of Europe. The question of pure food only went national with the great margarine debates of 1886.

An oleomargarine advertisement.

The Battle for Margarine

"All margarines are basically the same," wrote Professor Marion Nestle in 2006, "mixtures of soybean oil and food additives. Everything else is theatre and greasepaint." Nestle herself, an advocate of unprocessed foods, prefers to eat a little butter or olive oil, because "why would you want to put soybean oil on your bread?"[50]

Still, millions of people do; or palm oil, or sunflower oil, as the case may be. In 1997, more than twice as much margarine as butter was consumed per head in the United States (4.2 pounds of butter, 8.6 of margarine).[51] Some eat margarine because it is less expensive than butter; others because some particular brand seems to offer some promise of eternal health, or at least the promise of shedding a few pounds or lowering cholesterol by a few points (these elixirs may well be more expensive than butter). There are those who see margarine as a cheap substitute; others view it as a superior alternative. Some just eat it out of necessity because they cannot digest the dairy content of butter. In every case, however, the salient point about margarine for the consumers who choose to eat it is that it is *not* butter, and is not pretending to be. And even those who don't eat margarine are perfectly happy to tolerate its presence on the supermarket shelves, because they feel confident that it is not being marketed under false pretences.

Are we right to be so tolerant? What if the mere existence of margarine were itself a kind of fraud, perpetrated both against consumers and against butter? In Quebec, the only kind of margarine you can buy is uncoloured, so that if the substance melting on your toast is white, you can be sure it isn't butter. In 2000, the food giant Unilever tried and failed to have this Quebecois restriction overturned.[52] The same was true until quite recently in various parts of the United States. Until 1967, the state of Wisconsin, the Dairy State, influenced by the powerful butter lobby, banned the sale of yellow margarine, on the grounds that such margarine was ipso facto a swindle. (By Wisconsin standards, this law was comparatively moderate; a state law in place from 1925 to 1927 had banned the manufacture, sale, and even possession of margarine in any form whatsoever.)[53] In the 1950s, the only margarine you could buy in Wisconsin was pure white, so that no one could conceivably mistake it for a natural dairy product. When the housewife bought her white block of margarine, she would also be given a small sachet of yellow food colouring, which she could mix in when she got home, if she wanted to give the margarine a more buttery look, without being deceived about what she was doing. But for some this wasn't enough—as Wisconsin lawyer Barry Levenson discovered, "longtime Wisconsin residents remember the 1950s and early 1960s for bootleg runs to Illinois for yellow margarine."[54] Think about it: crossing the state line to get a genuinely fake version of a fake product so as not to be reduced to the indignity of having to fake it for yourself.

The twentieth-century Wisconsin ban on yellow margarine was a hangover from 1886, the year when margarine was the subject of fierce debates in the U.S. Congress concerning both its novelty and its inauthenticity—a "greasy counterfeit," as some called it. One congressman went so far as to name margarine "*the* monumental fraud of the nineteenth century." Another expressed bewilderment at the way that cunning trickery could compound a substance "to counterfeit an article of food—it is made to look like something it is not; to taste and smell like something it is not; to sell like something it is not; and so [to] deceive."[55]

Much of the initial American hostility to margarine stemmed from the circumstances under which it had been invented. Margarine was suspicious because it was French and associated with poverty. In 1886, Senator Palmer complained that margarine dated from the siege of Paris of 1870, a desperate time when "house pets were sold in the market for food."[56] Margarine had actually been patented the year before, in 1869. In the 1860s, there had been a shortage of edible fats in Europe. Napoleon III sought a cheaper version of butter. A French chemist, Hippolyte Mège-Mouriès, found the answer—a new method for emulsifying beef suet. First he turned the suet into oil by "digesting" it with chopped up cattle stomachs; then he emulsified this oil with the chopped up udders of cows, plus some bicarb of soda. Mège-Mouriès found the resulting cow-fat spread so precious—so opalescent—that he named it oleomargarine, after the Greek for pearl (*margaron*).

To the dairy farmers of America, however, the oily new substance was considerably less attractive than its name would suggest. The first U.S. patent for making margarine was granted on 30 December 1873. In the 1870s, U.S. agriculture was suffering a depression. and the influx of cheap butter substitutes added insult to the dairymen's sense of economic injury. Many of the oleomargarines made using the Mège patent were blatantly pretending to be butter. They were dyed a deep daffodil yellow and packed in the same tubs that butter traditionally came in. Whether they were made from hog fat or beef fat or a mixture, they rejoiced in the delightful name of "butterine." In 1877, New York and Pennsylvania passed laws requiring honest labelling, but this did little to stem the inexorable rise of margarine. In 1880, the United States exported nearly 40 million pounds of butter and only 20 million pounds of margarine; by 1885, the figures had reversed, with only 21.5 million pounds of butter exported and 38 million pounds of margarine. The total production of "oleo" was somewhere in the region of 50 million pounds by 1885.[57]

Mark Twain gives a glimpse of how gung-ho and unscrupulous some of the margarine sellers were. On a Mississippi riverboat headed for Cincinnati, he overheard a salesman who specialized in selling "butter" that was really margarine, talking to another salesman over a late breakfast. Twain saw the first seller take out a slab of "ostensible butter" and exclaim:

> You can't tell it from butter; by George, an expert can't! It's from our house. We supply most of the boats in the West; there's hardly a pound of butter on one of them. . . . You are going to see the day pretty soon, when you can't find an ounce of butter to bless yourself with, in any hotel in the Mississippi and Ohio valleys, outside of the big cities. . . . And we can sell it so dirt-cheap that the whole country has got to take it. . . . Butter's had its day—and from this out, butter goes to the wall.[58]

It is hardly surprising that dairy farmers should have seen margarine as the enemy.

When "oleomargarine" was debated in the House of Representatives and Senate in the spring and summer of 1886, Representative Robert M. LaFollette of Wisconsin called it "a villainous device for making money lawlessly, and subtilely eating the heart out of an industry which is to the Government what blood is to the body."[59] Margarine was an attack on agriculture, which meant it was an attack on America itself. Other congressmen from dairy-producing states concurred. The New York representative complained that the dairy industry was "in danger of extinction" unless the national legislature could protect it; state laws had been no use.

As well as being a threat to agriculture, margarine was depicted as a menace to public health. Dairymen insisted it was a "poison."[60] The Illinois representative, John R. Thomas, argued that not just farmers but consumers should be saved from margarine by the federal law, since they would not save themselves: "'It would seem that Americans delight in being defrauded and that they go yawping around with their mouths open, seeking to be taken in."[61] Various representatives insisted that margarine was rancid, or made from "filthy" fats taken from diseased animals or those fed on the swill from distilleries (they had learned how effective this charge could be in triggering public unease). The *Washington Post* referred to margarine makers as "manipulators of condemned suet and spoiled soap grease."[62] A congressman from Virginia depicted margarine as "the stomachs of pigs, sheep and calves reduced by acids, and then bromo-chloralum used to destroy the smell and prevent detection of the putrid mass.... Yes, 'cheap food' in the form of an apothecary's shop in the poor man's stomach."[63]

Some of its detractors thought that margarine had the power to attack not just the body but the soul as well. There were suggestions that margarine was ungodly, since it baffled "the four senses which God has given us." Being such a "promiscuous" mixture, perhaps it would lead to moral promiscuity in those who ate it. Margarine represented a lower way of life than butter. Congressman William Grout of Vermont argued that to resort to eating margarine would be a sign that the course of American civilization was reversed, with its citizens having moved backwards from being butter-eaters to eating the "raw tallow and lard which were the delight of our Saxon ancestors in the forests of Germany."[64]

Margarine, however, had its friends in Washington too, and they fought the increasingly hysterical tone of the butter lobbyists in a measured and reasonable tone. Butter's allies had called margarine a "midnight assassin," but "where was there a single tombstone engraved dead from oleomargarine?" asked Thomas Browne in Congress. As for claims that bad or "filthy" fat was used, a scientist pointed out that unless the fat used for margarine was fresh, it was "utterly worthless." Anyway, butter itself was not as bucolic and healthy as the farmers made out; it had become an industrial food in its own right, with all the risks that implied. A Pennsylvania congressman observed that butter was often adulterated with carrots, beets, and potatoes. There were also cases of rancid butter being "recovered" or deodorized before being sold on to unwitting consumers.

The crux of the debate was not whether oleo was poisonous but whether it was deceitful. The dairymen laboured the point that this "bogus butter" was often passed off as the real thing, a basic swindle since the cost of producing it was so much less than the cost of producing butter.[65] In 1879 it was said that consumers almost "invariably buy it for butter," rather than for what it really was.[66] Farmers testified that suspicious new "creameries" had opened up, which took in only small amounts of milk each day, and yet somehow managed to churn out vast quantities of "butter."[67] The reason? It was really "oleo butter," which was making "exorbitant profits" for its swindling manufacturers. A good deal of "oleo butter" seems to have found its way into boarding houses and restaurants, where customers asked for "butter" with their meals and paid the high price of butter but were given oleo. "He Got Oleomargarine Though He Asked for Butter at a Railroad Restaurant" was one pedantic headline in the *New York Times* in December 1886.[68]

It was easy to agree that this fraudulent sale of oleomargarine was wrong. But did it follow that no oleo should be sold? The *Washington Post*, fed up with the "incessant and wearisome wail" of the butter men, argued that it was fine to sell oleo, so long as it was "plainly labelled" as such with the name of the manufacturer and a government stamp on the package.[69] Consumers would know that butter was butter and oleo was oleo. It would be obvious that the latter was not butter because of its low price. Equally, butter's expense would be a test of its authenticity.

The butter men—all too predictably—disagreed. They insisted that so long as margarine was allowed to be coloured butter-yellow, it would continue to be passed off as butter, or mistaken for it. In testimony before the Senate Committee on Agriculture in 1886, Senator

Palmer, for the butter lobby, argued that the "rich, buttercup hue" of butter was sanctioned by the Bible, and that it was therefore wrong to dye margarine yellow with annatto.[70] Senator Henry Blair of New Hampshire made much the same point: "You may take all the other colours of the rainbow, but let butter have its pre-empted colour." Margarine's friends pointed out, however, that it was hard to say exactly what butter's "pre-empted colour" really was. Depending on the time of year and what the cows ate, butter itself varied widely in hue, sometimes turning a deep yellow, other times tending to a pale off-white. If it was fraudulent to add yellow colouring, then it was a fraud that the butter manufacturers were themselves guilty of, since much butter was itself now dyed yellow to meet consumer expectations. A Mr. Hinshaw, one of the new breed of Chicago margarine exporters, told the Senate that the margarine-makers actually had a prior claim on the yellow colouring, since the butter-makers had probably got the idea of using yellow dye from them; and that it would be only just for the margarine-makers to demand that the dairymen use a different colour from yellow. Let them have the other colours of the rainbow instead!

In the end, though, the butter lobby was too powerful. The federal Margarine Act of 1886 imposed a tax of two cents per pound on

An American advertisement for margarine, referring to
the fact that it cannot be bought coloured yellow.

margarine—less than the original bill, which called for a crippling ten cents a pound, but still punitive given that margarine needed to sell more cheaply than butter (which then cost about fourteen cents a pound).[71] The Margarine Act also called for manufacturers and wholesalers of margarine to buy expensive licences. This law was still not enough for the dairymen, though, who went back to the states to secure even more protection. New Hampshire, whose senator had wanted butter to keep its God-given colour, passed a law requiring all margarine to be coloured a bright pink. Minnesota, another dairy state, followed suit.[72] These bizarre "Pink Laws" were overturned in 1898 by the Supreme Court on the grounds that they were so extreme they were effectively a ban on margarine, and therefore "unconstitutional."[73] The legality of margarine had become a question of high constitutional politics, which is what both sides wanted. Yet neither side was satisfied with the outcome.

Harvey Washington Wiley

The temporary triumph of butter over margarine in 1886 was not a triumph for pure food over fake, but a victory for sectional interests and scaremongering over rational debate. The role of government in the whole affair, as a soapbox for the different sides rather than an instrument for bringing them together, did not inspire much confidence for the future. By the final decade of the nineteenth century, food politics in the United States still seemed hopelessly divided between the fanatics and the partisan interests, with national government caught fatally in the middle. But as so often in this story, it took one person to make the difference. He was a man of a distinctive type: extrovert but also somewhat priggish, a zealot for reform but not a fanatic for change, someone who understood the worlds of both science and big business without being in hock to either. He was, like Accum and Hassall, a man with a healthy sense of his own importance, as comes through in the autobiography he wrote at the end of his life to make clear just how important he had been in changing the face not just of food but of politics in the United States. Like Accum, he was a bit of a showman, but he had a great deal of

Hassall's remorseless determination as well. He believed himself to be a man of destiny, and though, as with so many such men, the self-importance sometimes grates, it was crucial to what he was able to achieve. What the cause of pure food needed at the end of the nineteenth century was a new, independent-minded champion, neither crank nor commercialist, but possessing a streak of both. Harvey Washington Wiley was this person.

A U.S. postage stamp depicting Harvey Washington Wiley (1844–1930), who did so much to secure the Pure Food and Drugs Act of 1906.

An intelligent, tall, well-built man, "muscled like a ditch digger," Wiley was born in a log cabin in rural Indiana in 1844.[74] His father was a farmer and a lay preacher, who was known as the "original abolitionist of Jefferson County." The whole family was educated to be bitterly antislavery. Wiley Jr. was brought up on the story of how, in 1840, his father had turned up at a polling station to vote for Martin van Buren, the "Free Soil" antislavery candidate, the only vote cast for him in the entire township. In those days, voting was oral, so there was no way of hiding which way you inclined. As Wiley Sr. marched up to the polls, he was assaulted with jeers of "Nigger! Nigger!" from a boorish and threatening crowd. This only strengthened his determination to do what he felt was right, and he never let young Harvey forget that this was the Wiley way.[75]

The son lost his father's religion, but he kept his ability to hold fast to unpopular views, even in the face of ridicule; indeed, unlike his father, he seemed actively to relish making himself ridiculous. After serving briefly on the Union side in the American Civil War, Wiley studied medicine, first at Indianapolis, then at Harvard. In 1874, he became one of the first faculty members at the newly founded Purdue University in Lafayette, Indiana. He would spend the next nine years on and off teaching there, but he made himself deeply unpopular with the board of trustees by flouting convention. Wiley liked to ride a bicycle around town—a high-wheel velocipede with one tiny wheel at the front and a giant wheel at the back, said to be the first bicycle ever seen in Tippecanoe County—wearing a pair of outrageous knickerbockers, frightening not just the horses but the local dignitaries as well. One of them compared him to "a monkey on a cart wheel."[76] It was not how a member of faculty ought to behave. The trustees also disliked the fact that he played baseball with the students and neglected to attend morning prayer. "In short, I was irreligious, frivolous and undignified," Wiley would later tell an interviewer, with a characteristic air of self-satisfaction.[77]

Despite the clowning, it was during his time at Purdue that Wiley began to take a serious interest in the battle against food adulteration. In 1878, he travelled to Germany, where he attended the lectures of the distinguished chemist August Willhem von Hofmann (1818–1892) on chemistry and Wittmack on adulteration.[78] He was elected to the German Chemical Society and began to work analysing sugar chemistry. Wiley had always been interested in sweetness, having learned how to make maple syrup as a boy from the 125 trees in his family's maple grove. Then, in the late 1850s, sorghum was introduced into southern Indiana, a grasslike plant whose canes yielded a sweet juice. The Wileys began to grow this too from seed, turning it into dark sorghum syrup after it was harvested.[79] After the Civil War, when the supply of New Orleans sugar and molasses was cut off, this home-grown sorghum supplied the Wiley family with all the sugar they needed. Wiley drew on these early memories in the 1870s when he started to analyse various different sugars using a "polariscope" that he had obtained in Germany—an instrument that used rays of light passed through a prism to test the chemical makeup of different sugar solutions. In testing sugars, he was motivated in part by the dream that America itself might become self-sufficient in sugar, just as his own family had been, rather than spending a hundred million dollars each year on imported cane sugar.[80] Wiley started by analysing maple and sorghum syrups, the sweeteners of his childhood, before moving on to a new and controversial source of sugar: glucose.

Like margarine, glucose—made in the United States from Indian corn—was a relatively recent creation, and many people were equally willing to denounce it as a sinister novelty. George Angell railed against glucose as a terrifying "giant" that had "grown in a few years to colossal proportions."[81] The question of size was undeniable; by the late 1870s, glucose was a two-million-dollar industry.[82] But Angell also insisted that glucose was bad for health, and this point was much less clear-cut. Wiley himself had mixed feelings about glucose. Essentially, he welcomed it as a wholesome, home-grown food, writing in the *Popular Science Monthly* in 1881 that "Corn, the new American king, now supplies us with bread, meat and sugar, as well as the whisky which we could do without."[83] On the other hand, his studies of various sugars showed him just how widely glucose was used not in its own right but as an adulterant. In the same article in 1881, he gave the example of honey. Much liquid "honey" at this time was really cleverly disguised glucose, with just a hint of the real stuff left in to give the right smell. Wiley knew that honey forgers went to great lengths to hide their deception. Sometimes, they went so far as to "put the remnants of bees, wings, legs etc. to carry out the fraud" so that the unnatural cleanness of glucose would not give its true nature away.[84] Shameless swindlers would take an artificial comb, fill it up with glucose, and cap the cells with paraffin.[85] According to the same article, "some inventive Yankee," so Wiley wrote, had gone so far as to patent this devious process.

When beekeepers read this, however, they did not see the funny side, and Wiley's remarks were interpreted as a slur on the integrity of honey. Honey journals referred to the notion of an artificial glucose comb as "the Wiley lie." The fear was that the whole honey industry would be damaged by the charge, and Wiley recalled that "I was made the object of bitter attack by beekeepers all over the US for years." Wiley did his best, though, to assure the beekeepers that he was on their side; that his intention was not to encourage adulteration but to prevent it. He wrote a report to the Indiana State Board of Health examining the sugars and syrups on sale, including honey, and the extent to which they had been adulterated. Once the beekeepers saw Wiley's evidence of "the vast injury done to the honey industry" by its adulteration with glucose, they came round to his side and, as he happily recorded in his autobiography, some of them "became my most enthusiastic supporters."[86]

Beekeepers were not the only ones to have taken note of Wiley's sugar analyses. While attending a conference of sorghum growers in St. Louis, Wiley met Dr. George Loring, the U.S. Commissioner of Agriculture, and made a good impression on him. In April 1883, the call came from Washington for Wiley to become the chief chemist in the Department of Agriculture. Wiley happily accepted, leaving the provincial scholars of Purdue for the intense political activity of the capital. At first, he continued his work on sugar, pressing for the establishment of a domestic sugar beet industry, against opposition from various interest groups. Later, he would claim for himself the title of "father of the beet-sugar industry."[87] Increasingly, though, he devoted himself to studying adulteration wherever he could find it, using his lab for the systematic analysis of various foods, bringing out reports on dairy products (1887); spices and condiments (1887); fermented alcoholic beverages (1887); lard (1889); baking powders (1889); sugars (1892); tea, coffee, and cocoa (1892); and canned vegetables (1893). He also spoke out against the "wretched and disgraceful evil" of adulteration in medicine, attacking patent medicines and the "various nostrums, salves, appliances, poisons, magic and sheer fraud this group of ghouls foisted upon . . . suffering humanity."[88] He was equally incensed by false advertisements for food, which claimed that special foods could "feed the brain, or feed the nerves, or feed the skin."[89] In 1892, he was elected president of the American Chemical Association. He was well on the way to establishing himself as the dominant figure in the fight against adulteration in the States.

There were others in Washington besides Wiley who interested themselves in pure food at this time. In 1889, Senator A. S. Paddock of Nebraska had brought a pure food bill before Congress, hoping to capitalize on some of Wiley's work (and on the margarine wars of a couple of years previously). But it failed; the bill was ridiculed, and then thwarted, as were subsequent, similar bills during the following decade. But Wiley was not downhearted. He understood the importance of choosing the right battles, and he had a peculiar knack for garnering the most effective kind of publicity for his cause; he also had an unusual hunger for the fight. In his youth, he had considered a career as a prizefighter, and he never lost his fighting instincts. There was an inevitable degree of self-aggrandizement in his struggle, and recent historians have shown that he was as concerned as any other Washington insider with increasing the power and prestige of his own federal bureau in relation to other government agencies.[90] But Wiley retained a clear sense of who his real enemies were—those who used chemicals to preserve food; dishonest labellers; manufacturers of adulterants; and sellers of patent medicines—and a pugnacious desire to defeat them. He knew that these interest groups would smugly label anyone who attacked them as "cranks" or "reformers without much business sense."[91] And he was determined not to let them get away with it.

It also helped that ridicule was something the thick-skinned Wiley knew how to deal with. To a certain degree, he guarded himself against it, fighting his cause in as reasonable and uncrankish a way as possible. Compared to other pure food campaigners, his language was not messianic, but humorous and commonsensical. He made a point of avoiding exaggeration, observing that the extent of adulteration was often overstated, and that it affected perhaps 5 percent of the food supply. On many questions he was neither a purist nor a prohibitionist. "It is not for me to tell my neighbour what he should eat, what he shall drink, what his religion shall be, or what his politics. These are matters which I think every man should be left to settle for himself. . . . Anything under heaven that I may be pleased to do I want the privilege of doing, even if it is eating limburger cheese" (a famously smelly cheese).[92]

But when he found something he did want to attack, he went after it with every sinew in his vast frame, daring the mockers to scoff. Fake honey and bleached flour; ineffectual headache powders and ludicrous panaceas; concealed caffeine and "rectified" or artificial whisky; these were Wiley's great bugbears. His most famous battle, though, was against manufacturers who claimed to be able to stave off the march of time itself: the makers of the new breed of preservatives.

Preservatives and the Poison Squad

"There is but one proper way to preserve foods," Wiley told a lecture audience in Washington, DC, in 1897, "and that is through sterilization and afterward hermetically sealing the vessels in which they are kept."[93] Most manufacturers took a different approach. With the twentieth century looming, an array of new wonder products was now at their disposal to arrest the inevitable tendency of food to decays. Manufacturers could now buy Freezine or Freezem, Rosaline or Preservaline to delay the decomposition of milk, cream, ice cream, sausages, hamburger steak, fish, and bulk oysters, or "almost every conceivable food that will spoil."[94] Wiley talked in his lecture about the growing use of salicylic acid, which "paralyses organic action" but also has a "deleterious effect upon the digestion."[95] The canning industry, which was growing by the day, made considerable use of additives: saccharine to make corn sweeter, copper to make peas greener, and all sorts of preservatives to stop meat from going off. Wiley had an instinctive dislike of these additives, which he felt could only deceive the consumer.

When presented with the argument that these new preservatives were no different from older ones, such as sugar, salt, and spice, Wiley would vehemently disagree. Spice advertised its presence with its distinctive taste, whereas these new chemicals were mostly devoid of taste or smell, so people who ate them had no idea they were there. Old-fashioned preservatives were, in Wiley's word, 'condimental', adding a great deal more to the food than the mere fact that they helped it to keep, whereas these new additives were merely 'chemical." To prove his point, he asked why little salt cellars of borax (otherwise known as sodium borate, a widely available industrial preservative described by Wiley as a "germicide" for "paralyzing fermentative action") should not be placed on the table along with the salt and pepper, so that people could add it or not as they chose.[96]

In 1898–99, the whole nation woke up to the potential problem of preservatives when the "embalmed meat" scandal broke.[97] During the Spanish-American War of 1898, American troops in Puerto Rico and Cuba sent frequent complaints home about the quality of the beef they were fed, both canned and so-called fresh—"putrid," "pungent," "stringy," and "gristly" were some of the terms used. During previous wars, such as the Civil War, animals were taken with the armies and slaughtered as needed. By the 1890s, however, America had a far more efficient meat-packing and processing industry, and it was decided that the army should be supplied with refrigerated beef, along with canned beef. Both kinds of meat drew complaints from soldiers, and these coincided with a growing incidence of disease in army camps.

This was a war with few direct American casualties in the line of battle but many incidental deaths, and in September 1898 President McKinley set up the Dodge Commission to investigate the conduct of the War Department during the war. In December 1898, a Major General Nelson A. Miles testified that 337 tons of "so-called refrigerated beef, which you might call embalmed beef," had been sent to the troops in Puerto Rico over the summer.[98] It was the word "embalmed" that was so shocking, suggesting as it did a human corpse. Another witness, Major W. H. Daly, an army surgeon, testified that he had seen "fresh beef" on an army ship sailing from Puerto Rico that smelt like an embalmed human body and tasted like decomposed boric acid. He also testified that he saw a quarter of beef hanging in the sun for sixty hours that remained completely, suspiciously, untainted. Meanwhile, numerous reports were heard of fighting soldiers having been disgusted by their rations of canned beef. Often it made them feel so ill they would throw it away rather than eat it. To be alarmed by preservatives no longer marked you out as a crank.

Ironically, when asked by the U.S. government to investigate army meat, Wiley, the arch enemy of preservatives, now found that the trouble in this case was not preservatives at all. "Embalmed beef" was largely a fiction. Wiley and his assistant, W. D. Bigelow, analysed samples of army canned beef and found no trace of any chemicals used as preservatives except for good old-fashioned salt—no borax,

181

no boric acid, no sulfites, no salicylic acid, and no benzoic acid.[99] By contrast, when his bureau tested ordinary commercial samples of canned meats a few years later, they found chemical preservatives in 6 percent of samples. Wiley concluded that the reason the rations had made the soldiers feel so nauseated was twofold. First, the army diet was badly constructed—day after day of cold slimy canned beef, with no potatoes or rice to balance it. "The human stomach does not tolerate the recurrence of the same article of food daily very well ... just as a person can not eat a quail every day for thirty days in succession, although for one day, with a bottle of wine, it is a very palatable ration."[100] His second reason was that the meat spoiled quickly in the heat of Cuba and Puerto Rico, which explained the foul stench. What he did not say was that, unlike a fine dinner of quail, the canned army beef was of extremely poor quality, fatty, stringy, and coarse.

Even if the legend of "embalmed beef " was not true, in Wiley's view, it was still extremely useful for his crusade. With the scandal fresh in the public's mind, the time was right to focus on the use of preservatives in the civilian diet. The trouble was how to prove that chemical preservatives were harmful. Wiley himself was convinced they were, to a greater or lesser extent. At the Mason Hearings of 1899–1900, when a Senate committee investigated the extent of the adulteration of food and drink, Wiley as chief chemist testified that "there is no preservative which paralyzes the ferments which create decay, which does not at the same time paralyze to an equal degree the ferments that produce digestion."[101] In other words, such preservatives by their very nature must be bad for the stomach. Yet at the same hearings, a Chicago manufacturer of preservatives named Albert Heller testified that the opposite was true: far from being harmful, preservatives such as formaldehyde were healthy, since they prevented people from getting cholera from milk products. Their very artificiality marked them out as useful weapons in the battle against tainted food. For example, Heller insisted, boric acid was an altogether beneficial ingredient for curing bacon, which might otherwise spoil. "I wish to say that every one of us eats embalmed meat—and we know it and like it and continue to eat it."[102]

Evidently, if Wiley were to continue in his view that preservatives were generally harmful, he needed some hard evidence. In 1902, with the cause of pure food gathering ground, he came up with his most provocative idea yet—the series of experiments that would come to be known as the "poison squad," for which he was given five thousand dollars from Congress. In his autobiography, Wiley boasted of his belief in practical experimentation: "I believe in trying it 'on the dog.'"[103] In this case, it was not a dog but twelve robust young men from the Department of Agriculture, who volunteered to live in a "scientific boarding house" and follow a strict preservative-enriched diet prescribed by Wiley. The idea was as brilliant in its simplicity as *Supersize Me*, the film in which the director Morgan Spurlock subjected his body to a diet of nothing but McDonald's burgers and fries for a month to prove beyond doubt that such a regimen damages your health.

Harvey Wiley carefully weighing food to be given to his famous "poison squad."

Wiley's poison squad was very similar, except that, unlike Spurlock, he did not use his own body but those of his volunteers. Half were to be fed a normal, wholesome, preservative-free diet, with

183

plenty of fruit and vegetables and no alcohol, cooked by a civil service cook. The other half would supplement this diet with a ration of preservatives. Every aspect of their health would be strictly monitored. Each member of the squad was required to record his weight, temperature, and pulse rate before each meal, to note down meticulously what he had eaten, and to carry with him sample jars for his urine and faeces, to be sent off to the government chemists.[104] A rigidly exact regime had to be followed. The coffee drinkers among the squad had to stick to two cups a day, no more and no less, lest any variation in caffeine intake should skew the results of the trial; "if at any time he shall feel disinclined to partake of the soothing cup," said one instruction, "he will have to relegate his feelings to the rear and drink whether or no."[105]

The experiment caught the public imagination at once, in part thanks to vivid press reports, which viewed the whole business in essentially comic terms. The term "poison squad" was coined by a young reporter on the *Washington Post*, George Rothwell Brown. Wiley himself disliked the phrase, since it prejudged the question; the

Harvey Wiley dining in Washington with his "poison squad" of strapping young human guinea pigs.

experiment was designed to find out whether borax was or was not poisonous. But he could not complain about the publicity. Brown followed the progress of Wiley's inmates in absurdist detail, describing how Wiley's food weigher would go so far as to bite a bean in half to make sure that precise portion control was maintained.[106] Poems circulated, one of which, the "Song of the Pizen Squad" by S. W. Gillian, was quoted by Wiley himself:

> For we are the Pizen Squad.
> On Prussic acid we break our fast;
> We lunch on a morphine stew;
> We dine with a matchhead consommé,
> Drink carbolic acid brew;
>
> . . .
>
> Thus all the "deadlies" we double-dare
> To put us beneath the sod;
> We're death-immunes and we're proud as proud—
> Hooray for the Pizen Squad![107]

The first chemical to be tested, in November 1902, was borax.[108] Wiley chose it partly because it was widely used, and partly because it was, according to Wiley, the "least objectionable" preservative and he would therefore be putting his twelve young men in the least danger by starting with it. There was a political reason too: Germany, as part of a long-standing meat war, had banned imports of American foods containing borax as unsafe (a German experiment conducted on four men had concluded in June 1902 that borax was indeed unsafe, leading to weight loss).[109] Wiley's much larger-scale experiment was important for recovering American face; whatever the results might be, at least America could claim to have had the more thorough experiment. Wiley insisted at the outset that he was keeping an "open mind" about borax.

For weeks the experiment stalled, to the amusement of George Rothwell Brown in the *Washington Post*, because Wiley found it so hard to get his human guinea pigs to the right weight. "One Boarder Grows Too Fat and Another Too Lean" read the headline on 16 December 1902. Wiley was fixated on the idea that his volunteers must be strapping young fellows of "normal" weight, working on the assumption that if the borax did *them* harm, it would be so much more harmful to children, the elderly, or the vulnerable. But, to his despair, one of his boarders kept gaining weight—"that boy eats like two men," he complained—while another was so scared by the experience of being in the scientific boarding house that he lost all appetite and began to shrink away.[110] Even after the borax dosing finally started, Wiley's troubles did not end. He had decided to give the borax surreptitiously, so as not to affect the mental health of those eating it. Initially, he hid powdered borax in the butter, but the boys soon cottoned on to it and began to eat less butter. The same happened when he hid it in milk, meat, and coffee, so he was forced to change tack and administer the borax openly, in capsule form, halfway through the meal, or sprinkled throughout.[111] At last the borax regime had started.

Within a few days, those on the borax diet started to lose weight. George Rothwell Brown reported the misery of those in the boarding house as their borax diet started to bite, getting particular relish out of the poison squad's gloomy Christmas celebrations. On 25 December 1902, Wiley had left Washington to celebrate the holiday in Indiana, leaving his boarders in a maudlin state. "You can't have borax and Christmas at the same time," one of them told Rothwell Brown. Another described their "festive" meal, which went as follows:

Apple Sauce
Borax
Soup
Borax   Turkey   Borax
Borax
Canned String Beans
Sweet Potatoes   White Potatoes
Turnips
Borax
Chipped Beef Cream Gravy
Cranberry Sauce   Celery Pickles

Rice Pudding
Milk   Bread and Butter   Tea   Coffee
A Little Borax[112]

There was obviously an element of exaggeration here; still, life in the boarding house does not sound much fun. Chef Perry, the temperamental civil service cook, seems to have been constantly grumbling about the conditions, bothering Wiley for a pay increase because he used to be chef to the queen of Bavaria and was not happy that "he puts borax in the skillet now instead of salt."[113] The inmates themselves, meanwhile, grew tired of the monotony of their existence. There were several cases of boarders stealing extra rations of food at night—a boiled egg here, a slice of bread there—something that Wiley took a very dim view of.[114]

Wiley's critics complained that it was not even clear what the point of this elaborate exercise was. "How does Professor Wiley propose to prove anything?" asked the *Washington Post*, complaining that by putting his poison squad on a special regime to build them up before the experiment began, Wiley was skewing the results; the real question was how borax affected "the average person." rather than these strapping youths.[115] An editorial in the *New York Evening Post* complained:

> Apparently it has not occurred to Dr. Wiley that experiments on such healthy robust young men will not be of much service. Nobody claims that borax is a violent and immediate poison, like arsenic or strychnine. It is merely believed or suspected that it interferes with digestion, and thus in the long run impairs the health of those whose stomachs are not vigorous. Young men in the student age do not usually know they have stomachs; they are apt to boast they could eat boiled brickbats on toast. Borax is not likely to affect them visibly; but children and adult dyspeptics—most American adults are dyspeptics—must be injured by a chemical which arrests fermentation, disguises the badness of tainted meat and fish, and retards digestion.[116]

Soon after these criticisms were levelled against Wiley, a "high official" at the Agricultural Department made the creepy suggestion of expanding the programme to include tests on invalids and infants, the "weak and sickly," so that the effects of borax on these more vulnerable groups could be gauged.[117] "Baby Class in Borax." squealed the *Washington Post*. Wiley himself disclaimed the idea, insisting that it had come from his boss, James Wilson, the secretary of agriculture, with whom he had difficult relations. Wiley did agree hypothetically, though, that using babies and the sick "would make the test more complete of course" and pondered, with somewhat sinister logic, that "as for obtaining the subjects, that wouldn't be hard. We could get the babies from infant asylums and foundling hospitals and there are plenty of invalids."[118]

An infant poison squad never materialized. Yet the mere fact that the idea of feeding preservatives to infants could be suggested by officials exposed something callous in the enterprise, a certain chilly disregard for the human guinea pigs. While some criticized the poison squad as ineffectual, other criticized it as irresponsible, putting the lives of these young men at risk. In 1903, Len Dockstader's minstrel show performed a song about the poison squad called "They'll Never Look the Same":

> If you ever visit the Smithsonian Institute
> Look out that Professor Wiley doesn't make you a recruit.
> He's got a lot of fellows there that tell him how they feel,—
> They take a bowl of poison every time they take a meal.
> For breakfast they get cyanide of liver, coffin shaped,
> For dinner they get undertaker's pie all trimmed with crepe;
> For supper, arsenic fritter, fried an appetizing shade,
> And late at night they get a prussic acid lemonade!
> O they may get over it but they'll never look the same.
> That kind of bill of fare would drive most men insane.[119]

Though Wiley denied that he ever allowed his experiments to be carried on "to the point of danger to health,"[120] this was disingenuous. By his own admission, he didn't know just how dangerous these preservatives were before the trials began; otherwise, why would he need to do the experiment? What's more, he confided to George Rothwell

Brown in 1903 that at times, "the dose has been as large as the men could stand."[121] During his subsequent benzoate trials of 1904, only three members of the squad continued to the end; the others became so ill, with inflamed oesophagi, extreme stomach pains, dizziness, and weight loss, that they had to withdraw.[122]

What Wiley ought to have said was that, for him, the end justified the means. At the end of a series of his poison squad trials, Wiley could pronounce what he called a "clear and unmistakable lesson": that "preservatives used in food are harmful to health."[123] If he had to give a few "young robust men" some stomachaches and sore throats in the process, so be it. The public soap opera of the "chemical café" had paid off.

In the following years, a number of states passed new and more effective pure-food laws, and finally, the dream of an omnibus federal law to regulate food standards across the country started to look not just realistic, but unavoidable. The House of Representatives twice passed a Pure Food and Drugs Act, though on both occasions the bill died in the Senate. Meanwhile, Wiley helped to keep the issue in the public's mind. He collaborated with Samuel Hopkins Adams, a journalist who in 1905 published a series of articles in *Collier's Weekly* attacking frauds in drugs: expensive fakes such as Liquozone, which marketed itself as a "universal antiseptic" but really consisted of 99 percent water.[124] Wiley also worked with the women of the National Consumers' League, which in 1905 announced the consumer's fundamental rights: to safe food, to truthful information, to choose, to be heard, and to be protected by government agencies.[125]

When at last the U.S. Pure Food and Drugs Act was passed, on 30 June 1906, Wiley felt triumphant. In his autobiography, he compared himself to a general who "wins a great battle and brings a final end to hostilities."[126] He complained that President Theodore Roosevelt had been given "undue credit" for this act of 1906. Yes, Roosevelt signed the bill into law, but he had not championed the act during its bitter fight for passage in Congress. He had not fought the good fight as Wiley had. The implication was that Wiley himself deserved sole credit for the new law. No one ever accused Harvey Washington Wiley of false modesty.

189

He was forgetting, or conveniently ignoring, someone else, though. If Wiley laid much of the groundwork for the 1906 law, the immediate impetus came from a much less probable source—an unheralded novel written by a nervous young socialist named Upton Sinclair. Perhaps unsurprisingly, Sinclair's name is absent from Wiley's autobiography. Wiley's part in the story is not yet finished; we will return to him later in this chapter. But the man of the hour in 1906 was Upton Sinclair. It was because of the unprecedented impact of his novel *The Jungle* that even the most diehard opponents of government intervention in the food business felt cowed into submission. Sinclair brought the public debate about food to such a pitch of anxiety and disgust that federal legislation was the only answer.

Upton Sinclair, Theodore Roosevelt, and The Jungle

The Jungle tells the story of Jurgis, a Lithuanian who comes to America with his family to be a "free man," and a rich one, but who finds himself instead living in the stockyard area of Chicago, enslaved to horrible and low-paying work as a meat packer. Jurgis's initial optimism turns to grim misery, as he works sweeping entrails off the gory floor. A century after the novel was published, the Zola-like descriptions of blood-splattered "killing beds" still have the power to shock. Sinclair makes you see both the nastiness of the "Packingtown swindles" and the squalor of forcing human workers to collude in them. The account of what goes into sausages was "an image that haunted the nation":[127]

> There was never the least attention paid to what was cut up for sausage; there would come all the way back from Europe old sausage that had been rejected, and that was mouldy and white—it would be dosed with borax and glycerine, and dumped into the hoppers, and made over again for home consumption. There would be meat that had tumbled out on the floor, in the dirt and sawdust, where the workers had tramped and spit uncounted billions of consumption germs. There would be meat stored in great piles in rooms; and the water from leaky roofs would drip over it, and thousands of rats would race about on

it. It was too dark in these storage places to see well, but a man could run his hand over these piles of meat and sweep off handfuls of the dried dung of rats. These rats were nuisances, and the packers would put poisoned bread out for them; they would die, and then rats, bread, and meat would go into the hoppers together. This is no fairy story and no joke; the meat would be shoveled into carts, and the man who did the shoveling would not trouble to lift out a rat even when he saw one—there were things that went into the sausage in comparison with which a poisoned rat was a tidbit.[128]

What made the image so haunting was that it wasn't fictitious. Sinclair himself had spent seven weeks observing the huge Chicago meat plants owned by rich "beef barons" and had seen such sausage making with his own eyes. He went to Chicago in 1904 after the failure of a Packingtown strike. Sinclair, then an impoverished and unhappily married writer and passionate socialist, aged just twenty-five, had written an article in favour of the strikers in a socialist newspaper, *Appeal to Reason*. The leaders of the strike read it and invited Sinclair to come to Chicago to witness their lives for himself. Because these plants put so little emphasis on human individuality, it wasn't hard for Sinclair to fit in. One worker told him that all he had to do was to carry a dinner pail and wear old clothes and everyone would take him for a working man. "You can see anything you want to see."[129]

What Sinclair witnessed were the most distressing and grotesque scenes of his life. This was no place for an anxious young man with stomach worries and a tendency to panic. Sinclair, who was the product of an alcoholic father and a puritanical mother, was said to have had "an obsessive fear of alcohol, sex and impurities of any kind." For days, he bravely persevered in the blood and the stench, "white-faced and thin," as he later described himself, but determined to expose the horrors he witnessed. He also interviewed those who knew the ways of Packingtown—workers, doctors, nurses, settlement-house workers—before retreating to a rural cabin to write the book. "I wrote with tears and anguish, pouring into the pages all the pain that life had meant to me."[130] In one of the few things he had in common

191

QUICK JURGIS WE MUST RECOVER THE BODY FROM THE LARD VAT

ALL STAR FEATURE CORP. PRESENTS IN MOTION PICTURES
— UPTON SINCLAIR'S —
WONDERFUL STORY OF THE BEEF PACKING INDUSTRY
THE JUNGLE
FEATURING
GEORGE NASH - GAIL KANE
AND THE AUTHOR
5 DARING ACTS — 210 ASTOUNDING SCENES

A poster for the film version of Upton Sinclair's *The Jungle*.

with Wiley, Sinclair also saw the fight against adulteration as a continuation of the previous generation's fight against slavery; several prominent readers compared *The Jungle* to *Uncle Tom's Cabin* for its political impact.

Sinclair was not writing with any particular axe to grind regarding the American diet. It was only later that he became a figurehead of the wilder fringes of the vegetarian movement. Aged twenty-six, he was still eating "white flour and sugar and other denatured foods" and suffered rotting teeth and a fragile digestion on account of it.[131] What he lacked in nutritional knowledge he made up for in an acute sensitivity to misery and gore. Five publishers rejected *The Jungle* on account of its shocking details—much too much "blood and guts," said a reader for Macmillan; publishers may have also been put off by the fact that the novel had previously been serialised in the social- ist *Appeal to Reason*. Doubleday took a risk and agreed to publish it, and ultimately it was the shocking details that made *The Jungle* such a success. The prose may be overwrought and the characterization of the Lithuanian workers somewhat wooden, but it is still, as Jack London said, a great book because it is "brutal with life."[132]

Sinclair depicts workers falling into vats and no one bothering to fish them out, so that "all but the bones of them had gone out to the world as Durham's Pure Leaf Lard"; meat inspectors turning a blind eye to diseased meat; old women with "ghastly pallor" "twisting sausage- links and racing with death"; "potted chicken" that is really rotten pork. Sinclair's message is that these individual scandals are not isolated but part of a giant industrial conspiracy: "The great corporation which employed you lied to you, and lied to the whole country; from top to bottom, it was nothing but one gigantic lie."[133] *The Jungle* is about a system of food production in which adulteration is the only currency. Jurgis and his family are well versed in the spectrum of spoiled-meat swindles that go on in Packingtown.

> Jonas had told them how the meat that was taken out of pickle would often be found sour, and how they would rub it up with soda to take away the smell, and sell it to be eaten on free-lunch counters; also of all the miracles of chemistry which they performed, giving to any sort of meat, fresh or salted, whole or chopped, any color and any flavor and any odor they chose. In the pickling of hams they had an ingenious apparatus, by which they saved time and increased the capacity of the plant—a machine consisting of a hollow needle attached to a pump; by plunging this needle into the meat and working with his foot, a man could fill a ham with pickle in a few seconds. And yet, in spite of this, there would be hams found spoiled, some of them with an odor so bad that a man could hardly bear to be in the room with them. To pump into these the packers had a second and much stronger pickle which destroyed the odor—a process known to the workers as "giving them

Hams being doctored in Upton Sinclair's *The Jungle*.

thirty per cent." Also, after the hams had been smoked, there would be found some that had gone to the bad. Formerly these had been sold as "Number Three Grade," but later on some ingenious person had hit upon a new device, and now they would extract the bone, about which the bad part generally lay, and insert in the hole a white-hot iron. After this invention there was no longer Number One, Two, and Three Grade—there was only Number One Grade.[134]

But the fact that they are aware of these tricks cannot save the workers from being duped by other food swindles in their turn.

How could they know that the pale blue milk that they bought around the corner was watered, and doctored with formaldehyde besides? ... How could they find out that their tea and coffee, their sugar and flour, had been doctored; that their canned peas had been coloured with copper salts, and their fruit jams with aniline dyes? And even if they had known it, what good would it have done them, Since there was no place within miles of them where any other sort was to be had? ... If they paid higher prices they might get frills and fanciness, or be cheated; but genuine quality they could not obtain for love or money.[135]

Because the problem was so systematic, Sinclair believed that the solution must be systematic too: a wholesale transfer from capitalism to socialism. In a rather clunking ending, which reviewers of *The Jungle* found unconvincing and which even Sinclair himself admitted was poorly written, Jurgis finds a vision of a better future by converting to socialism. "We shall bear down the opposition, we shall sweep it before us. . . . CHICAGO WILL BE OURS!"[136]

Except, of course, that it never was; the socialist revolution Sinclair dreamed of failed to materialize, least of all in Chicago, whose gangster days were only just beginning. What his book did provoke, however, was an instant outcry over the disgusting and deceitful way in which Packingtown meat was produced. The revelations in *The Jungle* were terrifying for a nation where per capita meat consumption was as high as 179 pounds a year, nearly half a pound of meat per person a day.[137] Sinclair himself was disappointed. He had hoped to awaken in his reader a sympathy with the plight of the workers; instead, the main response had been a selfish, if understandable, bourgeois fear of being poisoned by tubercular meat. "I aimed at the public's heart and by accident I hit it in the stomach," was his famous quip.[138]

One of the stomachs hit by *The Jungle* was the sizeable paunch of President Theodore "Teddy" Roosevelt, who had been sent an advance copy by Sinclair's publisher, Doubleday.[139] Roosevelt had been interested in the meat supply ever since he fought in Cuba during the Spanish-American War at the time of the "embalmed beef" crisis. He remembered how his men had started complaining about the quality of the canned meat. Catching one of his soldiers throwing his rations away he accused him, with typical masculine bluntness, of being a baby. "Eat it and be a man," he barked. But when the soldier tried to obey Roosevelt's order, he vomited. Roosevelt then tried to eat some of the meat himself, but "found I could not . . . [it] was slimy, stringy and coarse . . . like a bundle of fibres."[140]

When the assassination of William McKinley propelled Roosevelt into the White House in 1901, he revisited beef. Roosevelt's domestic policy rested on the so-called square deal, a promise to curb the power of monopolistic corporations or "trusts" within "reasonable limits." Roosevelt became known as a "trustbuster," issuing forty-four lawsuits against overweening corporations in the course of his presidency. Several years before *The Jungle* appeared, the beef trust was already in Roosevelt's sights, as one of the most powerful commercial oligarchies—or "oligopolies." In 1902, there had been wide public protest when the price of meat—which was controlled by the Beef Trust—shot up. Over six months, the price of sirloin steak had risen from eighteen cents a pound to twenty-two cents; of shoulder of lamb, from eight cents to twelve cents; of pork chops, from twelve cents to fifteen cents. These increases were extremely unpopular with retail meat dealers as well as with customers, the increase in the wholesale price of meat forcing some of them out of business. "Poor people cannot stand the increases, but the retailer has to make them," commented a meat-seller at New York's Washington Market.[141] These price hikes were especially scandalous, because they coincided with a collapse in the prices paid by the big meat packers at cattle markets.

The packers, it seemed, were engaged in outrageous profiteering. The Justice Department investigated, and in May 1902 Roosevelt ordered the attorney general to bring a bill against the meat packers in Chicago, accusing them of colluding to suppress competition in the sale of livestock and to fix the retail price of meat.[142] The legal process was painfully slow; it would be 1905 before a judgement was reached. Roosevelt tried to hurry along his antimonopoly work by setting up a Bureau of Corporations in 1903 under James Garfield and getting it to investigate the Beef Trust; but to his disappointment, Garfield—who seems to have been "entertained too well at the Saddle and Sirloin Club in Packingtown" while making his inquiries—concluded that beef prices were "reasonable."[143]

Meanwhile, the Beef Trust grew in power. In a series of articles for *Everybody's Magazine* in August 1905, Charles Edward Russell called it "The Greatest Trust in the World" and complained that the Big Four Packers had a power greater "than in the history of men has been exercised by king, emperor, or irresponsible oligarchy."[144] Russell claimed that, by suppressing competition in the cattle market, the packers had caused ruin of many farmers and the failure of thirty-two banks.[145]

When *The Jungle* arrived on Roosevelt's desk in early 1906, it provided the opportunity he had been waiting for. Only two years remained of Roosevelt's second term as president, and if he was going to curtail the Beef Trust, he would have to act fast. *The Jungle* was the weapon he needed, but given its provenance it had to be carefully handled. It had certainly created the necessary public alarm to push through political change, selling 25,000 copies in just six weeks (and 100,000 by September). The trouble with *The Jungle*, from Roosevelt's point of view, was twofold. First, he had to ascertain whether the shocking revelations it contained were true. Second, he had to do his best to ignore Upton Sinclair's emotional brand of politics.

On 15 March 1906, Roosevelt wrote to Sinclair from Washington telling him that he had read "if not all, yet a good deal of your book" and asking him to visit in the first week of April. He then spent the bulk of the letter admonishing Sinclair for the foolishness of his so-cialism. "Personally, I think that one of the chief early effects of such attempt to put socialism . . . into practice, would be the elimination by starvation . . . of that same community on whose behalf socialism would be invoked." He agreed with Sinclair that "radical action must be taken" against "the effects of arrogant and selfish greed on the part of the capitalist" but could not bear what he perceived as Sinclair's hysteria. "A quarter of a century's hard work over what I may call politico-sociological problems had made me distrust men of hys-terical temperament," he insisted. Only at the very end of the letter, in a handwritten postscript, did he offer Sinclair the assurance he needed—"that the specific evils you point out shall, if their existence be proved, and if I have power, be eradicated."[146]

Sinclair went to Washington, as invited. He ate luncheon at the White House with Roosevelt and some members of his "tennis cabi-net." Sinclair found the president both appalling and amusing, and he was very struck by the way that he clenched his fist and shouted in-discreet abuse about his enemies, hitting "the table at every accented syllable." Roosevelt for his part seemed to have been confirmed in his opinion of Sinclair as "hysterical, unbalanced and untruthful." But no matter. This odd couple were together in the fight against the pack-ers now. As Roosevelt wrote sometime later to a friend, "I could not afford to disregard ugly things that had been found out simply be-cause I did not like the man who had helped in finding them out."[147]

Roosevelt appointed two commissioners to go to Chicago and investigate the allegations in *The Jungle*. The commissioners—Charles P. Neill and James Bronson Reynolds—asked Sinclair to go with them, but he refused, sending two socialist friends in his place, Ella Reeve and Richard Bloor.[148] Ella Reeve confirmed that "no words are adequate to paint the horrors of the packing houses."[149] In addition, she reported to Sinclair her fears that the official Neill–Reynolds report would be a whitewash. The commissioners were supposed to be operating in secret, but the packers soon rumbled them. On 10 April, the *Chicago Tribune*—a newspaper, needless to say, extremely favourable to the packers—ran the headline "PRESIDENT HUNTS IN 'THE JUNGLE,'" claiming that Roosevelt's commissioners had "failed to verify a single important statement made in the book." It reported that the president was planning to "pillory the author of the book by holding him up to public disapproval."[150] Another article, the next day, claimed that *The Jungle* was "95 per cent lies."[151] In a state of high distress, Sinclair sent the president two letters and two telegrams protesting that Roosevelt had allowed "falsehoods to be telegraphed to the Chicago Tribune." Roosevelt replied with typical bluff calmness,

A cartoon depicting Theodore Roosevelt and the great Meat Scandal of 1906.

assuring Sinclair that there "has been no official whitewash or official anything sent out from Washington," adding, in a postscript: "Really, Mr. Sinclair, you *must* keep your head."[152]

Sure enough, when Neill and Reynolds did report back orally to the president in May on conditions in the Chicago stockyards, they confirmed most of the revolting truth of *The Jungle*. The only detail they could not corroborate, as Sinclair recalled, was "of men falling into vats and being rendered into pure leaf lard. There had been several cases, but always the packers had seen to it that the widows were returned to the old country."[153] Roosevelt and Sinclair held different opinions on what should be done with the report. Sinclair wanted it published. Roosevelt argued that going public would be pandering to "the apostles of sensationalism" and would damage honest ranchers and farmers "who have been guilty of no misconduct whatsoever" by lowering wholesale meat prices.[154] Besides, holding the report back could be a useful political bargaining chip for Roosevelt. In May 1906, his colleague Senator Beveridge had introduced a meat inspection bill, and Roosevelt hoped to use the threat of releasing the report as a way of bringing the packers into line: it was "a club to force the passage of the bill." Sinclair disagreed: "genuine reform" would only come from "enlightened public opinion."[155]

In the end, they both had their way. Sinclair used the press to put pressure on Roosevelt to release the Neill–Reynolds report.[156] And Roosevelt got his meat inspection bill, though not without a fight in Congress from meat interests. On 26 May, the Beveridge Meat Inspection Bill passed the Senate, requiring the inspection of all "cattle, sheep, swine and goats slaughtered for human consumption." All carcasses found unfit to eat, and all artificially dyed meat, were to be destroyed. Anyone attempting to bribe inspectors could be fined up to ten thousand dollars. "The rendering of all fats into lard" was to be "closely supervised by employees of the Department"; no more human beings were to be turned into lard, was the subtext.[157] The *Washington Post* reported that "Fifteen minutes before it was passed not a Senator would have admitted that the bill had a ghost of a chance to become a law—certainly not this session. Its passage is the direct consequence of the disclosures made in Upton

Sinclair's novel, 'The Jungle.'"[158] On 30 June, Roosevelt signed it into law.

For Sinclair himself, the Beveridge bill was hopelessly inadequate, "like plugging up one leak in a dam, and making a devil of a fuss about it and letting a dozen other leaks go unnoticed." What was needed, he thought, was not just inspection but municipal slaughterhouses owned by the government, as in Europe, where the slaughterhouses were "as cleanly as modern hospitals, and not to be compared in any way with the filthy shambles we endure here."[159] This was typical Sinclair. Municipal slaughterhouses implied a radically different politics to that of Roosevelt's America—a fundamentally different vision of how far the state should intervene in public health. Originally, municipal slaughterhouses were one of Napoleon's great ideas, although the first public abattoirs in Paris only opened after his fall, in 1818.[160] In Britain, municipal slaughterhouses sprang up in various cities— including Manchester and Glasgow—after the Public Health Act of 1875 and were associated with mild socialism of the Fabian variety:[161] not the sort of thing that would recommend them to Congress.

Roosevelt, ever pragmatic, was more satisfied with the Beveridge bill. The meat inspection law was, in his words, an "important" remedy to the "scandalous abuses" of Packingtown. In his own *Autobiography*, he observed an interesting after-effect. "The big beef men bitterly opposed" the new law, yet three or four years later, "every honest man" in the beef business was in favour of the bill, discovering that it actually helped their business concerns rather than harming them.[162] Tougher inspection, for those willing to accept it, was a route to greater revenue, not less. This would also be true of the new rules governing the labelling of food contained in the Pure Food and Drugs Act, brought into law that same summer's day.

Honest Labels and Pure Ketchup

It was said of Harvey Washington Wiley that he was so unyielding in his hatred of mislabelled food that he would even object to lady finger biscuits unless it could be proved that they actually contained amputated female digits. Something of the same spirit animated the

Pure Food Act of 1906. It was a law that placed a great emphasis on accurate labels. Twice as many regulations dealt with misbranding as with adulteration (fourteen as against seven).[163] Primarily, these labelling clauses were there to protect the consumer. Regulation 17 forbade "false or misleading" labels; regulation 25 banned the substitution of a "recognized" substance for another unless it was stated on the label; regulation 26 insisted that when "refuse materials, fragments or trimmings" were used, they must be declared as such on the label. All of these requirements conformed to Wiley's definition of purity in food. "I never used the word 'purity' except in one sense. A pure food is what it is represented to be."[164]

There was, however, another dimension to these labelling regulations. As well as protecting the consumer, they sought to regulate trade practice; two aims that were not necessarily the same. Regulations 20, 21, and 27 dealt with products with "distinctive names," stating that "a 'distinctive name' is a trade, arbitrary or fancy name which clearly

A commemorative stamp marking the anniversary of the Pure Food and Drugs Act of 1906.

distinguishes a food product, mixture or compound from any other food product, mixture or compound."[165] It was now illegal to sell counterfeit versions of these "distinctive name" products. This was no longer simply a question of selling margarine as "genuine butter" or potted pork as "chicken"; no longer a question of purity, in Wiley's sense. It was about protecting the manufacturer's trademark.

Trademarks in the modern meaning only came into being in the 1870s. The idea was to give goods a brand equivalent to the old guild marks—a mark of quality that would offer assurance to the consumer and instil brand loyalty. Trademarks are designed to make you, the consumer, feel safe. Your feeling of safety matters to the manufacturer not for its own sake, though, but because it will make you buy more of the product, and possibly pay a higher price for it. Branded goods are there to make you feel you are not being swindled. Yet this in itself can become a kind of scam, as happened in the United States after 1906.

Though its ostensible purpose was to make food labels more accurate, the Pure Food and Drugs Act also made possible a new dimension to trademarking. To license a trademark in the United States, permission had to come from the patent commissioner, in the Department of Commerce. If you wanted to market Aunt Jemima's pancake flour, say, or Campbell's soup, or Brer Rabbit molasses, you needed to register the label with the Patent and Trademark Office. Hundreds of packers used the opportunity provided by the new law to apply to the patent commissioner to register labels containing the phrase "guaranteed under the Pure Food and Drug Act, June 30th 1906," as if the government itself had overseen the creation of the food in the factories. Finally, in November 1908, the then commissioner of patents cracked down on this. From now on, he announced, he would license no more labels containing this phrase. The new law, he insisted, was not to be seen as a "warranty for purity."[166]

But many food retailers had tried to do just that: use the new law as a warranty for purity, and thus turn it to profit. Wiley represented it as the triumph of government and science against the swindlers of big business, but the case was more complicated than this, as he well knew. It was true that those business interests whose livelihood was threatened by the pure food agenda protested loudly in the run-up to June 1906. Chief among these were the food processors who used chemical preservatives—such as fruit dryers, who relied on sulphur dioxide, or sellers of dried codfish, who coated their product with borax—and the whiskey rectifiers. As Wiley said, "rectifier" was a misnomer, since far from rectifying anything, the whiskey rectifiers actually made "crooked whiskey" from raw alcohol mixed with flavourings, colours, and "ageing oils" to make a very nasty substance look like fine old whiskey.[167] In 1906, the rectifiers controlled 85–90 percent of all distilled spirits. Wiley was not himself a whiskey drinker, but he hated the deception involved. "A man is entitled . . . to get the character, quality and kind of material he asks for, and if you ask for whiskey you ought not to get a bottle of slush."[168] The congressman from Kentucky (a state where real whiskey was still made) agreed with Wiley, complaining that "the rectified stuff will eat the intestines out of a coyote."[169] It was inevitable that the whiskey rectifiers should have fought against the new law; and their fight paid off in part, given that the final bill, rather than banning rectified whiskey, only required it to be plainly labelled "compound," "imitation," or "blend."[170]

Overall, however, there were probably more commercial interests who came out strongly in support of the bill than opposed it. "CANNERS FOR PURE FOOD read a headline in the New York Times on 16 February 1907, announcing that the National Association of Canners and Packers had declared in favour of the law.[171] Once the bill was a reality, many in the food business recognized that, far from being threatened by it, they could make good money out of it. Even the swindlers of Packingtown spotted an opportunity. "PURE FOOD CENTRE OF THE WORLD," trumpeted a long article in the Chicago Daily Tribune on 25 February 1907, the same paper that had slated Upton Sinclair only a year earlier. This has led some historians to see the bill as fundamentally corrupt—as a story of "business control over politics," as Gabriel Kolko put it. For Kolko, the 1906 act was just an opportunity for the big food manufacturers trying to get a competitive advantage over their rivals.[172] On this interpretation, the so-called progressives of the Roosevelt era were really conservatives in disguise, willing to protect big commercial interests at all costs. This is going too far. There were plenty of officials in Roosevelt's government who recognized the need for better protection for consumers. Robert Allen, a government chemist, argued that better labelling of food was necessary because "when purchasers know where a product was made, when it was made and who made it, and are informed of the true nature and substance of the article offered for consumption, it is almost impossible to impose upon [even] the most ignorant and careless consumers."[173]

But the new consumer awareness of pure food could certainly cut both ways. Not only might consumer interests be opposed to business interests, business might also exploit the need to allay consumer anxieties. This was shown most clearly in the case of one of the most famous of all American food brands, Heinz tomato ketchup. Around the turn of the twentieth century, almost all commercial ketchups were preserved with benzoic acid, a colourless and odourless antiseptic that is found in nature in cranberries. Nineteenth-century formulas for ketchup were thinnish and tomatoey, with much less sugar and vinegar than modern recipes. Ketchup manufacturers argued that a small amount of benzoic acid was necessary to stop the ketchup from going off after it was opened. G. F. Mason, for example, who worked as the manager of Henry J. Heinz's research laboratory, found that when he tried to make benzoate-free ketchup, it blew the cork out of the bottle.[174]

Doubts were beginning to be raised, though, about the safety of benzoates. In 1904, Wiley had tested benzoates on his poison squad and found symptoms ranging from sore throats to dizziness, weight loss, and severe stomach pains.[175] Defenders of benzoic acid said that it was safe in small quantities—after all, the Lord himself put it in cranberries! Wiley was not convinced. The fact that the Lord put it in cranberries was neither here nor there. Cranberries were inedible in their natural state; anyway, "no chemist can ever imitate nature's combinations."[176] Wiley disliked benzoic acid not just because it was harmful in itself but because it could be used to make ketchup out of inferior-quality tomatoes swept off the floor.

The question still remained, though, as to whether good commercial ketchup could be made without benzoates. Most ketchup manufacturers claimed that it was impossible. When they tried to make their existing recipes without the benzoate, the ketchup fermented soon after it was opened. Some of the fault for this, argued Wiley sternly, lay with sloppy consumers, who did not look after their ketchup bottles properly.

> The habit of leaving a tomato ketchup bottle upon the table where the material adheres to the rim and becomes hardened to a gummy paste, serving as a pabulum for flies, does not appeal with any great force to the aesthetic sense relative to dining rooms. A ketchup bottle carefully opened and used in such a way as to avoid infection and then returned to the ice box can be kept for many days without danger of fermentation.[177]

It seems a little harsh to place all the responsibility for preserving ketchup on the harmless consumer. And what if you didn't have an ice box?

An anonymous critic of Wiley's wrote in the *Food Law Bulletin* that it was unfair of Wiley to rule against the use of benzoate-enhanced ketchup until he had informed manufacturers "how to put it up" without the preservatives.[178] Wiley, never one to refuse a challenge, announced that he would indeed be working to discover a method to make a safe, scientific, benzoate-free ketchup. He set two researchers on to the job, Avril and Katherine Bitting, who spent the summer of 1907 visiting more than twenty ketchup factories and stirring up countless batches of experimental nonpreservative ketchup at Loudon's ketchup factory in Terre Haute, using only the ripest, reddest tomatoes (since it was one of Wiley's convictions that benzoates were often used to disguise inferior tomatoes) and paying scrupulous attention to hygiene. They found that if they upped the amount of vinegar and sugar in the ketchup mix, it was possible to make delicious unbenzoated ketchup. Avril Bitting published their findings in 1909 in an article snappily entitled "Experiments on the Spoilage of Tomato Ketchup."

By the time the article appeared, however, the argument was becoming academic. No one could doubt that it was possible to make commercial ketchup without benzoates, because Henry J. Heinz of Pittsburgh had been doing exactly that since 1905. Before Wiley set to work on benzoates, Heinz ketchup had been no more nor less polluted than any ketchup. At various points, Heinz ketchup was coloured with coal-tar derivatives and preserved with either salicylic or benzoic acid. The manager of Heinz's research lab, G. F. Mason, insisted that there was no satisfactory way of making benzoate-free ketchup—consumers would always prefer the "nice, clean looking preserves" made with benzoates to the unreliable, mucky, easily fermented kind made without.[179] Mason didn't see that there was anything wrong with this, in any case, since benzoate of soda was not, in his view, a poison. A Heinz executive Sebastian Mueller defended his company's use of artificial preservatives at a Pure Food Congress in St. Louis in September 1904. The use of such preservatives, Mueller pointed out, was "recognized and permitted in all civilized countries." In less than a year, Mueller completely changed his mind about this and suddenly converted to the notion that if Heinz could make all their foods without preservatives, it would "revolutionize" the manufacturing process.[180] During 1905, the Heinz company managed to make half of all their ketchup—a hefty 1.8 million bottles—without preservatives. After the Pure Food Act was passed in June 1906, all preservatives were removed from the company's ketchup (even though the new law did not actually ban benzoates), and the Heinz company never looked back.

The reasons for this rapid change of heart have been debated. Was it the result of a discovery about the best way to manufacture ketchup, or was it a discovery about the best way to reassure worried consumers and get them to open their wallets? The answer was probably both. Official histories of Heinz argue that the great Henry J. Heinz had always been devoted to pure foods, ever since he watched his mother bottle up her own fresh horseradish as a child, and that his antibenzoate stand was high-minded and honourable.[181] It is true that he became one of the most influential supporters of the "pure food" movement, particularly once he saw how well the new product was selling. It was said that Heinz "never really liked chemists," and that removing benzoates from his ketchup came as second nature—which does not fully explain why he allowed his factories to put the benzoates there in the first place.[182] Heinz's conversion to pure food around

An early Heinz Ketchup advertisement, boasting that the ketchup is free from benzoates.

the time that government legislation was beginning to bite looks just a little too convenient. Some have even wondered whether Wiley and Heinz were in cahoots, since they were undoubtedly friends, and Heinz bolstered Wiley's career by using his endorsement in a new advertising campaign for his improved benzoate-free ketchup.[183]

But this mutually rewarding friendship need not be seen as sinister, so much as inevitable: both men recognized something of themselves in the other, not least a willingness to compromise in order to get what they wanted. As with so many manufacturers before and since who

207

have taken a stand on the purity of their food, Heinz's motives were mixed, as were Wiley's. Robert C. Alberts, author of a biography of Heinz, called it "noble purpose compounded with self-interest"—the story of much of human progress.[184] Indeed, Henry J. Heinz proved that noble purpose could serve self-interest much better than a grubby attachment to preservatives. "He knew in the long run it would be for the good of the Food Business," recalled a colleague.[185]

The new, benzoate-free Heinz ketchup was, however, far more expensive than the old variety had been. As the historian of ketchup Andrew Smith writes:

> When other nationally produced benzoated ketchups retailed for ten to twelve cents, Heinz's ketchup retailed at twenty-five to thirty cents. Obviously, Heinz paid more for fresh, ripe, tomatoes, but the cost did not exceed 15 percent of the total retail price. The other costs—additional raw materials (spices, sugar, vinegar), labour, packaging (glass bottles, labels, wrapping paper, bags, and packing material), overheads and freight—were presumably similar to those of other manufacturers. Heinz's major new expense was the need to convince consumers to purchase higher costing preservative-free ketchup as opposed to the less expensive benzoated ketchup. Previously Heinz had expended almost no money on advertising ketchup. After the passage of the national pure food law, Heinz's ketchup advertisements in magazines and newspapers exceeded all other manufacturers' combined.[186]

"Preservatives must Go!" read one of Heinz's ads. "What Every Woman Should Know About Benzoate of Soda in Foods!" read another. The Heinz campaign was ingenious, if a little unvarnished by today's standards. It appealed simultaneously to housewives frightened that they were poisoning their children and to grocers frightened that they might find themselves on the wrong side of the law. "SHALL YOUR FOODS BE DRUGGED OR NOT? asked one headline.[187] "Dr Wiley Condemns Preservatives," warned an ad in the *American Grocer* in 1909. "When the order prohibiting their sale comes, how will you be prepared for it?"[188] Actually, the order did not come prohibiting it; after a protracted battle, manufacturers won the legal right to continue using benzoate of soda as a preservative. By this point, though, most manufacturers had followed Heinz in removing preservatives. Curtice brothers, whose "Blue Label Tomato Ketchup" continued to use benzoate, went into decline.[189]

There are considerable ironies in the way that Heinz ketchup built its empire on its status as "pure food." As Heinz's competitors observed at the time, the only way to take the benzoate out of tomato ketchup was by creating concoctions "overdosed with sugar and vinegar."[190] There was some truth in this. The new formula for Heinz ketchup contained twice as much sugar and vinegar as before, and also more salt. Those in the probenzoate lobby insisted that benzoate was needed to "retain the full and natural flavour of the tomato."[191] Without the benzoate, tomato ketchup became something thicker, sweeter, and, to some tastes, more cloying.

Heinz tried to insist that much of the new sweetness came from using ripe red tomatoes instead of unripe fruit or tomato waste. Harvey Wiley had asked: "Why should the poor man pay approximately the same price for a bottle of catsup made from the sweepings of a tomato factory when he could get a great deal more catsup for the same price in a pure state?" But in fact, as we have seen, the price was not the same. What is more surprising is that Wiley—so relentless in his pursuit of detail in other respects—did not seem to pay much attention to the fact that the food value of the new ketchup was not the same either, since it was so much more sugary than the old preservative-laced ketchups.

Wiley couldn't have foreseen the exponential rise in sugar consumption that would contribute to the vast rise in obesity in the West by the end of the twentieth century; but already in his lifetime, doctors were warning of the dangers of diabetes associated with too much sugar, and health experts were decrying the sugar addiction of America's children. In the end, Wiley had too much faith in "purity" as a catch-all solution to the problems of safe food. Just because benzoated ketchup was a less than perfect product, did that mean that nonbenzoated but sugary Heinz ketchup should qualify as a health-giving product? The same dilemmas have dogged the question of "pure food" ever since.

What made this all the stranger was that it was not as if Wiley was unaware that eating too much sugar was a bad thing. In one of his books on nutrition, Wiley tells the following story, which is more revealing than he can have intended:

Avoid the possibility of forming the "sugar-habit." I was proud, only yesterday, at a gathering of grown people and children, where cake was served, to have my three-year-old son come running to me when the cake was brought out, crying out in great joy, DADDY, I WON'T EAT THAT CAKE; IT ISN'T GOOD FOR ME." Only a few days before this, his mother, hearing the noise of a battle in the front yard, ran to the window and saw the young dietician astride the prostrate form of Dickey, his even-aged playmate, administering an assorted variety of Jess Willard "knockout drops." "What are you doing, Harvey?" she cried. "Dickey will eat candy, and it isn't good for him" was the response.[192]

In the years after 1906, Wiley came to resemble his own three-year-old son, grappling an unwilling nation to the ground to administer his medicine, so fixated on poisons that he had begun to miss the bigger picture.

Saccharine and Caffeine: The Aftermath of 1906

Following the passage of the Pure Food Act, Wiley's power at the Bureau of Chemistry waned. Professor F. L. Dunlap was appointed acting chief in Wiley's absence, which Wiley saw as a way of undermining his authority.[193] By 1908, newspaper articles were calling him a "hated man," who had been deemed "guilty of insubordination" by his bosses at the Department of Agriculture, especially Secretary Wilson, with whom he had never got on ("a consummate hypocrite" was what Wilson called him in private).[194] Things came to a head earlier that year when Wilson invited a delegate of food manufacturers to the White House representing the benzoate of soda and saccharine industries. Wiley came too. The meeting started off well, from Wiley's point of view. He convinced Roosevelt of the dangers of benzoate of soda, at which the president thumped the table with his fist and told the manufacturers: "You shall not put this substance in foods!"[195]

The trouble began when conversation turned to saccharine, a substance that portly Roosevelt regularly took for purposes of slimming. A fruit packer stood up to defend saccharine, telling the president that his firm saved four thousand dollars the previous year by using saccharine in canned sweet corn instead of sugar. Wiley interjected at this point:

> "Yes, Mr. President . . . and everybody who ate that corn thought they were eating sugar, whereas they were eating a substance which was highly injurious to health."
>
> When I said this, President Roosevelt turned upon me, purple with anger and with clenched fists, hissing through his teeth, and said, "You say saccharine is injurious to health? Why, Dr. Rixey gives it to me every day. Anybody who says saccharine is injurious to health is an idiot."[196]

This was the final straw for Wiley's already shaky relationship with Roosevelt, and his influence in Washington never recovered.

It did not help that Wiley's own instincts on pure food were deserting him. Where once he had managed to bridge the gulf between commerce and the cranks, now he was looking distinctly crankish himself. In 1911, he brought a suit against the Coca-Cola company. It is possible to imagine many grounds on which he might have attacked the drink, but the one he chose was that it was mislabelled, since it contained no cocaine and very little cola.[197] What it did contain was caffeine, a substance that was not mentioned on the label and which he considered "objectionable," by implication as objectionable as cocaine. "In England," he insisted, "I have seen women who, if they were denied their tea at four o'clock, would become almost wild." Wiley analysed Coca-Cola and found it "habit-forming and nerve-racking."[198] His case against the company began in March 1911, while Wiley was on honeymoon (he had married late in life a woman thirty years his junior).[199] He lost, and he retired from government service a year later. The rest of his career was spent at *Good Housekeeping* magazine, where he established the *Good Housekeeping* seal of approval, continuing to give frequent lectures praising public health as "our greatest national asset."[200]

For Wiley's eightieth birthday in 1924, a special menu was produced. It read:

> *Alum pickle, coppered.*
> *Borated Baked Bluefish.*
> *Roosevelt Asparagus, Saccharine Dressing.*
> *Renovated Cream Butter.*
> *All food colours used guaranteed to be*
> *non-certified coal-tar products.*[201]

Wiley had been a man ahead of his time, but the joke at his eightieth birthday party suggests that he was by now a creature of the past. These were the obsessions of an earlier age. The remainder of the twentieth century would show that, for all his heroic work, Wiley's stand for preservative-free and accurately labelled food was just the beginning.

5

MOCK GOSLINGS AND PEAR-NANAS

Today we don't have anything so crude as adulterants; we have
additives and improvers and nutrients.
      —Elizabeth David, *English Bread and Yeast Cookery* (1977)

George Orwell was a child during the First World War. When he looked back on his early life, he confessed that his chief memory was not of all the deaths, but of all the margarine. The butter shortages caused by the Great War meant that margarine switched from being the food of the poor to being a universally used substitute—even for privileged scholars at Eton, such as Orwell. "By 1917, the war had almost ceased to affect us, except through our stomachs."[1] Orwell, in his clear-eyed way, saw how food was increasingly being reduced to a kind of unfood. He abhorred this "century of mechanization," which had, he believed, more or less eliminated the "taste for decent food." Thanks to tinned food, "synthetic flavouring matters, etc., the palate is almost a dead organ."[2] In Orwell's novel *Coming Up for Air*, the hero George Bowling (who is reduced to eating fake foodstuffs such as fish frankfurters or artificial marmalade) complains: "Everything comes out of a carton or a tin, or it's hauled out of a refrigerator or squirted out of a tap or squeezed out of a tube."[3] Orwell himself lamented the decline of the English apple:

As you can see by looking at any greengrocer's shop, what the majority of English people mean by an apple is a lump of highly coloured cotton wool from America or Australia; they will devour these things, apparently with pleasure, and let the English apples rot under the trees. It is the shiny, standardized, machine-made look of the American apple that appeals to them; the superior taste of the English apple is something they simply do not notice. Or look at the factory-made, foil-wrapped cheeses and blended butter in any grocer's; look at the hideous rows of tins which usurp more and more of the space in any food shop, even a dairy; look at a sixpenny Swiss roll or a twopenny ice-cream; look at the filthy chemical byproduct that people will pour down their throats under the name of beer. Wherever you look you will see some stock machine-made article triumphing over the old-fashioned article that still tastes of something other than sawdust.[4]

Orwell wrote these words in 1937. About this, as about so much else, he was prophetic. The world of synthetic food that he described so potently was only just beginning. Orwell was right that the twentieth century would be the century in which fake foods became the norm. He was also right that while these fakes started off as shabby substitutes—the reviled margarine of the Great War—they soon became seen by many as superior alternatives to the real food they replaced. By the old standards, this squirted, cartonned, squeezed food would have counted as adulterated. But as the century progressed, tastes and norms changed. The "shiny" and the "standardized" came to be prized above the real.

Ersatz Food and Wartime Fakes

Orwell's experience of the privations of war was the British one: joyless, faintly nauseating, but nothing compared to what happened on the other side. It was Germany that became the laboratory for what fake foods people could stomach. They even came up with a new word for it: the experience of *ersatz* food became one of the central features of the German experience of the First World War. Soon after war began, the British navy cut off sea supplies to Germany through a blockade, which resulted in widespread hunger on the home front.

With shortages of all basic provisions, Germany was forced to be ingenious in the creation of new foods. Essentially, these were new versions of the age-old famine foods that peasants had used since ancient times to stave off death (see chapter 2). The difference was that they had a veneer of modernity and were bolstered with vigorous propaganda from the German state, which insisted against the instincts of every German diner that potato was really quite as filling as bread; and later, that swede was as nourishing as potato. Exhibitions were held all over the country demonstrating the huge new range of ersatz foods on offer, in an attempt to make a virtue of necessity. There were 837 certified varieties of substitute sausage. An American reporter noted that German shopkeepers, when selling an ersatz food, did not bother to "palm it off on you by telling you that it is just as good. He merely says: 'It is the only thing of

A German placard advertising ersatz coffee.

the kind that can be bought in Germany.'" In this sense, there was no swindle because "there's little real anything to be bought in Germany today."[5]

There were "eggs" made from maize and potatoes; ersatz "lamb chop" that was really rice; ersatz "steak" that was made from spinach, potatoes, nuts, and (the final indignity) ersatz egg. Substitute piled upon substitute. Even the glamorous cafés of Berlin served up *Ersatzpräparate*.[6] At the start of the war, coffee (which in any case was usually chicory mixed with sugar beet) became roasted nuts flavoured with coal tar; later, this became too expensive and was replaced with roasted beech nuts and acorns; but later still, during the "turnip winter" of 1916–17, every spare acorn was used to feed pigs, and "coffee" was made from carrots and turnips, to accompany a meal of turnip stew and turnip bread. As one despairing commentator wrote: "It's bad when surrogates must be used, worse when bad surrogates, rather than good ones must be bought. Ethel Cooper, who lived in Leipzig during the war, commented that she did not mind consuming rat; it was the rat substitute she couldn't bear.[7]

Some of the foods considered ersatz at the start of the war were later accepted with gratitude. Jam was one. In 1915 and 1916, jam was loathed, because it was urged by the government as a substitute for meat fat and butter. Bread and jam as a main course! To a population used to meat, this was a wretched thought. You could hardly feed a family with bread and jam. In August 1916 women rioted in Kattowitz, screaming "Bread! Bacon! Fat! Potatoes! Away with jam!"[8] Little by little, however, as deprivation increased, people became happy to dine on jam. It no longer seemed ersatz, but a substance in its own right.

The great problem was the shortage of meat and fat. Pork and *Butterbrot* were the staples of the German diet. The war saw "unpleasant scenes" in front of butcher's shops as housewives battled each other for the last morsel of pig. The meat shortage led to disgusting attempts to make ersatz fats, of which margarine was perhaps the least horrible. With lard and dripping in short supply, Germans wondered in desperation if fats could be rendered from other animals. There were experiments with producing fats from rats, mice, hamsters, crows, and even cockroaches. There was even a plan to extract protein from the wings of dragonflies.[9] These crazy schemes are not so far removed from the insane substitute foods that get cooked up in the trenches by Baldrick in *Blackadder Goes Forth*, such as coffee (mud) with milk (spit) and sugar (dandruff), or filet mignon in sauce béarnaise (dog turds in glue). The difference, of course, was that the German ersatz creations were real and were eaten not just in the trenches but in normal civilian homes.

Ersatz food was a concoction sanctioned by war, in which the most fanciful adulterations were not merely legal, but even encouraged, as a patriotic means of conserving resources. Ash could be dolled up in a nice packet and labelled "ersatz pepper." To drink an infusion of ground walnut shells and call it "coffee" was not a sign of madness but of good citizenship. A new illness sprang up: "substitute sickness," where consumers were made ill from a combination of hunger and eating dire surrogate foods, many of which included the "indigestible remains of animals." Ersatz food contributed to the air of insubstantiality in wartime Berlin, where everyone lived in a state of suspense waiting for it all to end. People started applying the term *ersatz* to everything, even themselves. An *Ersatzmensch* was a substitute person, a simulacrum of a human being, who was no more real than the *Ersatzbutter* they spread on their *Ersatzbrot*.

This culture of fakeness continued during the deprived years of Weimar and lingered on into the prewar Nazi regime, whose propaganda promised mouthwatering visions of mass consumption, but whose shops held empty shelves and ersatz junk. Sugar continued to be advocated as a replacement for fat. The production of the once loathed jam trebled between 1933 and 1937. People were forced to use ersatz soap, which did not wash properly. The German taste for coffee and cakes had to be abandoned: in 1938 all cream cakes were banned as wasteful. Coffee was available only intermittently. Goebbels was scornful of those consumers weak enough to complain of the coffee shortages—these "pathetic old biddies" as he delighted in called them. "In times when coffee is scarce, a decent person simply drinks less or stops drinking it altogether," was his considered view.[10]

For all its pretence of newness, the Nazi regime offered the same old fantastical substitutes. All that was new was the propaganda, which reached fresh heights of nonsense. At one point, the lemon supply dried up, and the regime announced that it would be replaced with rhubarb. Apart from a general acidity, it is hard to see what lemons and rhubarb have in common. You cannot squeeze a stick of rhubarb over a piece of fish; or put a slice of rhubarb in a drink; or make lemonade from it; or use its zest because it has none; or, for that matter, use rhubarb in any of the ways that you might use lemons. Never mind. The Nazi information machine announced the substitution of lemon with rhubarb as a triumph, because lemons were imported and rhubarb was not: "Only through the German soil are the finest vibrations transmitted to the blood . . . therefore, fare thee well, lemon, we need thee not. German rhubarb will replace thee altogether."[11] The continued reliance on ersatz foods brings home graphically how Hitler's Germany had never recovered from the horrors of the First World War.

By comparison with Germany, the British home front was relatively comfortable during the Great War, apart from the ubiquity of margarine. Rationing was only introduced in 1918. Even so, cooks did find themselves reduced to using humbler ingredients, and less of them, than they might otherwise have done. "My Tuesdays are meatless / My coffee is sweetless," as the popular song went. Cookbooks of the era are replete with "mock" dishes, peculiar doppelgängers of luxury goods. Mock food had been a feature of the British kitchen since the seventeenth century, the most famous being mock turtle soup (hence the "mock turtle" of *Alice in Wonderland*). Real turtle soup, using turtles from the West Indies, was prodigiously expensive. Serving it was a sign that yours was the most upper class of households. Eating "mock turtle soup" enabled middle-class homes to ape their "superiors."[12] It was made from a boiled calf 's head and seasoned with ham, sweet herbs, madeira, and cayenne, the same seasonings as for real turtle soup.[13] Almost certainly, it tasted nothing like real turtle, except for a certain glueyness of texture. But how would those eating it know, since they had never tasted the real thing?

A 1928 advertisement for Campbell's mock turtle soup—the most famous of all the mock foods.

The Great War saw a resurgence in mock foods. *Our Grandmothers' Recipes*, a cookbook of 1916 by Lady Algernon Percy, included recipes for Mock Meat Pie (made with haricot beans, bacon, and onion) and Poor Man's Goose, a dish of liver with sage and onions. Another cookbook of the time advises making leg of pork "to taste like turkey" (because turkey was a more prestigious meat than pork), mock oyster patties (which are really salsify in cream), mock hare (minced beef and pork), and mock crab sandwiches (shrimps and herring roes).[14] Reading these mock recipes, the obvious question is, why did they bother? No one eating them can have been fooled by these dishes; and anyway, the word "mock" prevented any deception from taking place. Often, the "mock" made little attempt to resemble the real. Food historian Colin Spencer has come across a Victorian recipe for "a boiled salad, made from potato, celery, brussels sprout and beetroot," which "with a superhuman effort of the imagination pretended to be a lobster salad."[15] Mock food is the adulteration of make-believe. Enthusiasts for mock dishes were like children making mud pies, pretending they are cakes with cherries on top, the

219

difference being that most children know better than to eat mud pies. But then most children don't have to keep up appearances.

Make-believe food became still more common during the Second World War, when fantasy at the table seemed a good way to maintain morale. There was a change, though, in that the taste of these creations became less and less important, with more emphasis placed on "visual effect."[16] Mock chops (grated potatoes, soya flour, and onions), mock cream (evaporated milk mixed with gelatine), mock toad-in-the-hole (made with no egg in the batter): all of these recipes put more energy into looking right than tasting right. Ambrose Heath, author of *More Kitchen Front Recipes*, gave a recipe for "mock fish," consisting of ground rice, milk, a little onion or leek, and a seasoning of anchovy essence, all cooked into a kind of polenta-ish porridge which is then moulded into fish fillet shapes, fried until golden, and served with parsley sauce. Mock apricot flan was made of carrots and plum jam; mock oyster soup from fish trimmings.[17] Then there were the "mock goslings" of Josephine Terry, which turned out to be nothing but apple slices in potato pastry, served with gravy.[18]

Almost certainly, these mock foods were more written about than made. It is a dangerous business, inferring what people ate from the evidence of recipe books. If cookbooks were a perfect record of what the general population ate, then Britain today would be a nation where everyone sat down at night to immaculate meals of Nigella fish pie and Jamie Oliver roast beef. We know this isn't true. Still, written recipes can tell us something—they can tell us what people aspired to eat. And a world in which anyone aspired to eat mock gosling is a world in which kitchen substitutes had become commonplace.

After rationing began in January 1940, the public was also exposed to a new phenomenon: powdered substitute foods. Jack Drummond at the Ministry of Food introduced numerous dehydrated foods to the public, in particular, dried milk and dried egg. One London girl remembered wartime ice cream as being "an appalling mess, yellow and lumpy like scrambled eggs, with gritty little lumps of ice embedded in it," which sounds as if it was made with both dried milk and dried egg.[19] Still, rationing was more or less democratic. It affected grand restaurants as much as individuals. Mario Gallati, the manager of The Ivy in Covent Garden, remembered the foul "mayonnaise" that he invented during the war from flour and water put in the mixer with vinegar, mustard, and a little powdered egg. "It made me shudder to serve it, but everyone took this kind of 'ersatz' food very much in their stride."[20] In the same spirit, the nation embraced spam fritters, jam made with carrot, and eggless cakes.

More of a problem to the British consumer were the natural foods brought in under rationing as replacements for scarce proteins. Horse meat and whale steak were regarded with revulsion, even though both are nourishing and wholesome foods, considered delicious in many cultures. Most notorious of all was the tinned snoek or bara-couta introduced after the war (in 1947) as a replacement for tinned sardines. As the historian Colin Spencer has pointed out, "One would think that at a time when the butter and meat ration had been cut, the bacon ration halved and after a winter and spring of catastrophe, a new tinned fish which was cheap and unrationed . . . would be greeted with delight."[21] The Ministry of Food bought millions of tins of the stuff and published eight recipes for snoek, the most celebrated being snoek piquante, involving spring onions and vinegar.[22] Still, the British regarded snoek with deep suspicion. By 1949, more than a third of the snoek stock was still unsold. The ministry was forced to sell it off quietly as cat food.

The snoek affair showed how fed up people were with having their food choices dictated by the central government. Nutritionally speaking, the British ate better during the war than ever before or since: more wholemeal bread and vegetables, fewer sweets, small but regular amounts of meat and fish. In 1946, the chief medical officer reported that the "vital statistics" of the nation for the war years had been "phenomenally good." Thanks in part to rations for children of orange juice, milk, and rosehip syrup and better diets for pregnant women, child mortality rates dropped, even taking into account the deaths of seven thousand children in the Blitz. There were many fewer anaemic women and children, and many more children had perfect sets of teeth. Rationally, people ought to have felt well fed after the war. But they didn't. They felt deprived and dreary, fed up with broken biscuits and endlessly making do. As the food writer

Marguerite Patten said, when people looked forward to victory, one of the things they were looking forward to was more variety in their food.[23]

Nevertheless, the postwar attitude to substitute foods in Britain remained confused. On the one hand, there was a yearning to escape the stifling yoke of the Ministry of Food and enjoy once again the joy of choosing real foods for oneself, especially foods that were rich or exotic or imported: cream, bananas, fresh tomatoes. On the other hand, the years of rationing had acclimatized many housewives to using substitutes that they would once have considered shoddy. When Elizabeth David returned home to England after the war (having spent much of it in Cairo, where she ate delicious spiced pilafs, olive oil, ripe apricots), she was struck by the contents of her friends' pantries. "Everyone else had hoards of things like powdered soups and packets of dehydrated egg to which they were conditioned."[24] Such things were now completely normal and not even seen as cheats (whereas Elizabeth David found them cheerless, heartless, and dismal compared to the Mediterranean food she had come to love). Instead of turning away from processed and substitute foods after the war, the British public began to eat more and more of them, for they offered the illusion of freedom at a low cost. In the 1960s, a sign was spotted outside a village shop selling fresh fruit: "Lovely Ripe Pears—Good as Tinned!"[25]

Imitation Foods in the United States

Across the Atlantic, despite the absence of the privation conditions experienced in Europe, ersatz foods were taking over the grocery store too, though not without a fight from the manufacturers of "real" foods. The legal position of "imitation foods" in the United States was starting to change. Until the middle of the twentieth century, Congress and the states had generally used the law to protect basic American agricultural products from competition from novel processed foods. The Oleomargarine Act of 1886 (see chapter 4) was a case in point: margarine was taxed to prevent it destroying the market in dairy butter, and also to prevent the consumer from being misled. Similarly, the "Filled Milk Act" of 1923 prohibited interstate commerce in any milk to which nondairy fats or oils had been added, on the grounds that the sale of such products "constitutes a fraud upon the public."[26] The idea behind these laws was a familiar one: that basic, traditional products needed to be safeguarded against newer, fabricated substitutes.

The market had changed during the twentieth century, though, especially during the Depression years. As one cynical consumer of mock goslings and pear-nanas said, "there never was a product made but some ~~gosh darn~~ fool could make it worse and sell it for less."[27] Substandard versions of regular goods posed a dilemma for government. Technically, they were not allowed to be sold under the 1906 act. But without them, much of the already hungry population would starve. A compromise was reached in 1931 in the so-called Canner's Amendment, under which substandard but wholesome canned goods, such as blemished, split, or underripe peaches and pears, were permitted to be sold, but only if they bore an off-putting black crepe label, announcing the low quality. This move was welcomed as "an immeasurable help" to the hard-pressed housewife.[28]

The question of imitation as opposed to substandard foods was harder to resolve. During the early 1930s, countless new foods had been produced, with often ingenious advertising. The market was flooded with cleverly marketed, new, cheap imitation products, most with "fanciful" names such as Salad Bouquet, an imitation vinegar, Peanut Spred (a peanut butter with only modest peanut content) and Bred-Spred, a kind of jam with a lot of pectin and not much fruit. Real homemade jam needs the pectin in the fruit to make it set, but new technology enabled manufacturers to use refined pectin. They could now achieve a set consistency using nothing but sugar, water, and colouring: a kind of fruitless jam, though the law did not allow it to be called that. The problem was that the 1906 Pure Food and Drugs Act did not prevent such worthless products being marketed because of the "distinctive name proviso." This permitted foods to be sold that would otherwise have been considered adulterations of traditional products. So long as Bred-Spred did not call itself "jam" or "jelly," it was entitled to contain as little fruit as it liked.

This was galling—both to the manufacturers of real jam, whose product looked artificially expensive next to the likes of Bred-Spred, and to the Food and Drug Administration (FDA), whose job of maintaining food quality seemed undermined by these fabricated products. How was the ordinary shopper expected to know what was meant by their fancy names and gleaming packaging? At the height of the Depression, the president, Franklin Delano Roosevelt, complained that "the various qualities of goods require a discrimination which is not at the command of consumers," adding that "they are likely to confuse outward appearance with inward integrity."[29] Integrity was very important to Roosevelt. His New Deal programme of economic relief depended on hope, and hope was eroded by dishonesty. Roosevelt's White House was famous for its lacklustre, but always frugal and healthy food, overseen by his formidable wife Eleanor.[30] In a message to Congress in 1935, Roosevelt insisted that "honesty ought to be the best policy, not only for one individual or one enterprise, but for every individual and every enterprise in the nation."[31] Honesty, said Roosevelt, required that "the strict exclusion from our markets of harmful or adulterated products" and "the careful enforcement" of food standards.

In these Depression years, there was waning public confidence that harmful products really were being excluded from the market. In 1933, a best-selling book was published by a pair of consumer advocates, Arthur Kallet and F.J. Schlink, suggesting that the American population was essentially just *100 000 000 Guinea Pigs*, on which new food and drugs were constantly being tested, without any real guarantee that they were not poisonous. The book was reprinted thirteen times during the first six months after publication. Kallet and Schlink painted a scary scenario in which nothing was what it seemed and hidden poisons lurked in every package in the grocery store and pharmacy. The fact that the poisons were not all immediate in their effects—like the Victorian red lead and copper—made them no less sinister. "If the poison is such," they wrote, "that it acts slowly and insidiously, perhaps over a long period of years . . . then we poor consumers must be test animals all our lives." It is a chilling image. As for the food on the shelves, much of it was not really food at all, in Kallet and Schlink's view. Packaged pineapple pie, they argued, should for honesty's sake be relabelled as "corn starch-filled, glucose-sweetened pie made with sub-standard canned pineapple, artificial (citric acid) lemon flavour and artificial coal tar color."[32]

There was a gathering sense that without basic standards stipulating the quality of individual foods, the market would descend into anarchy. Hence, the 1938 Food, Drug and Cosmetic Act, signed into law by FDR on 25 June, was designed to mark a significant improvement of the 1906 act signed by Roosevelt's fifth cousin Theodore.[33] The new law was a triumph for honesty. As well as setting far more rigid regulations for drugs companies (in the wake of a scandal in 1937 when more than a hundred people, many of them children, died after taking a new "sulfa" wonder drug, which was really antifreeze), it clarified the federal government's power to police the food supply. The 1938 act finally abolished the distinctive name proviso and began to establish basic food standards. Now, it was a requirement for the label of a food to bear "its common or usual name" alongside whatever fanciful brand name it might choose. Bred-Spred could no longer deny that it was purporting to be jam. Gradually, the FDA also established enforceable standards for some of the most important foods of the American larder. The first standards were set for tomato products, to be followed by milk and cream, fruit juices, canned tuna, canned vegetables, chocolate, flour and grains, cereals, mayonnaise, and macaroni.

One of the earliest foods to be given a standard was jam or "jelly." Essentially, the new food standards were like old-fashioned recipes, laying down exactly what ingredients should be in any given food, and how much of each; this way, any swindling or tampering would be easily apparent, and also easy to punish. In the case of jam, the FDA took evidence from family recipes and cookbooks and decided that the correct recipe was about fifty-fifty fruit to sugar, a generous ratio. The legal minimum fruit content for jam was then set at 45 percent (compare this with the current UK regulation of 35 percent for most fruits). Assuming it is well made, a jam containing 45 percent fruit is the proper article—luscious and fruity, as well as being far more nutritious than the Bred-Spred horrors of the Depression era. By setting the percentage so high, the U.S. government was ensuring that "jam" would become worthy of the name.

For most of the 1940s, the new food standards worked well. They went some way to protecting the diet of a population hit by war and rationing (albeit on nothing like the same scale as in Britain). The government used the standards to protect consumers against swindles they could otherwise do nothing about. The recipe approach was one that everyone—consumers as well as lawmakers—could understand, and it was upheld by the courts. In 1944, there was a series of cases prohibiting substitute dairy products. The Supreme Court stated at the time that it was necessary to prohibit such products on occasion because labelling was not an adequate remedy for deception.[34] In 1949, Quaker Oats was banned from marketing a farina cereal with vitamin D, because it did not meet the statutory requirements for such products. There were cracks around the edges of the system, though. In some areas, demanding standards could lead to bootlegging. A Philadelphia lawyer insisted that he had seen bootlegged consignments of inferior ice cream travelling the city in refrigerated vans.[35] More significantly, the manufacturers of imitation foods also began to fight back.

The landmark case was the Imitation Jam ruling of 1952. In 1951, the FDA seized sixty-two cases of "Delicious Brand Imitation Jam" in assorted flavours. The "jam" was seized in New Mexico, where it had been shipped from Denver, Colorado; under the law, "interstate commerce" in adulterated goods was forbidden. The FDA ruled that this pseudo-jam was "misbranded" under the Food and Drugs Act, since it contained a scant 25 percent fruit (the rest being highly gelatinized sugary water). Under section 403(g), a food was misbranded if "it purports to be" a food for which a standard has been set. In the FDA's view, this product was clearly purporting to be jam, and, equally clearly, it could not be jam because it fell short of the fruit requirements by a full 20 percent. To the FDA, Delicious Brand Imitation Jam was a deception, despite the apparent honesty of its name. Even if the label admitted that it was not real jam as defined by the law, most consumers would still believe that it was.

The FDA, however, was deploying this case not only to uphold the purity of jam against imitators, but also to reinforce its own powers

to set food standards. The strategy backfired. After conflicting judgements were given in the trial court and the court of appeals, the
case went to the Supreme Court, which ruled that Delicious Brand
Imitation jam was *not* guilty of misbranding. Section 403(c) of the
Food Act stated that imitation foods were misbranded *unless* they
were labelled as such; but since the Delicious Brand preserves were
prominently labelled as "imitation," it did not matter that their fruit
content was so woefully low. The Supreme Court thus rejected the
view of the FDA that products such as Delicious Brand Imitation
Jam were inherently deceptive. As one supporter of the ruling said:
"There is nothing difficult or strange about the word imitation."[36]
After 1952, it would still be an offence to pass off imitation jam as
pure jam, but bakers would henceforth be free to sell "cake roll with
imitation jelly," and customers could buy an imitation jelly and peanut butter sandwich in a diner, probably for a lower price than if the
jelly had been the real, full-fruit kind.

The Imitation Jam case made a lot of people very worried. A lawyer for the association of National Milk Producers complained that
it would give rise to an explosion of substandard foods—not just
jams and jellies, but dairy products too. Consumers would become
confused about what the true standards were. "Unscrupulous" sellers
would take the opportunity "to deceive and cheat them."[37] What was
the point of having food standards if decent producers couldn't be
protected against poor imitations? Another American lawyer complained that imitation foods were not even as cheap as they seemed.
They pretended to be offering good value to cash-strapped consumers, but actually they had "an unfair price bulge when competing
with the genuine product":[38] in other words, they cost much more
than they were worth. The price of the imitation food was largely
determined by the price of the genuine article. The imitation version
would be priced up a few cents below the real thing. But the real
value of these imitation goods was often much lower still. The consumer was thus being cheated twice over: once, by buying the imitation food in the first place; and again, by buying it at a far higher
price than it warranted. Yet again, it all boiled down to consumer
knowledge, or lack of it. In 1894, the Supreme Court had stated that

"The Constitution of the United States does not secure to anyone the privilege of defrauding the public."[39] But in the eyes of its critics, the Imitation Jam ruling had done just that.

These were voices in the wilderness, trying to stem the tide of history. Soon, the idea that legislators could ban imitation foods would start to look quaint. In 1952, the year of the Imitation Jam ruling, America was poised to enter what has been called its "golden age of food processing."[40] Frozen orange juice, instant coffee, readymade TV dinners of chicken à la king, boil-in-the-bag macaroni and cheese in glitzy foil pouches, dehydrated potato salad—all these wonder products and many more were at the disposal of the housewife in 1950s' America. If you browse the commercial pages of America's regional newspapers for 1952, you find job advertisements for frozen food salesmen, special offers on Miracle Whip, promotions for Campbell's Tomato Soup and Hormel's Chili in a can.[41] The critical difference between these new processed foods and the older ones was that they had lost their inferiority complex in relation to unprocessed food. Paul Willis, a food industry boss, boasted in 1956 that "Today's processed foods have a food value at least equal, and often superior to, raw produce, but many housewives are still spending countless hours preparing raw produce in the erroneous belief that they are feeding their families more 'healthfully.'"[42] This marked a crucial shift. Thanks to intensive marketing, substitute foods were no longer to be seen as poor relations of the foods they originally mimicked. They were new; and new was best.

Additives, New Foods, and the White House Conference of 1969

In 1953, Dwight "Ike" Eisenhower became president. Eisenhower would be remembered for building the interstate highway system, and for his doctrine of "dynamic conservatism." As it applied to food, this meant embracing the new array of processed foods, and freeing food manufacturers from too much government regulation. It was farewell to the self-abnegation and wholesomeness of the war years and hello to fast cars on the open highway and as much "modern" food as you could buy. The new food industry had a lot in common with the ever-growing auto industry. Both were constantly seeking ways of adding "value" to their products and making them seem as exciting as possible; consumer safety was a secondary consideration. Soon after he assumed the presidency, Eisenhower attended a special "research" luncheon hosted by the U.S. Department of Agriculture in Beltsville, Maryland. The meal was designed to showcase the marvellous potential in all the new ways of processing foods. Eisenhower sampled "powdered orange juice, potato chip bars, a whey cheese spread, 'dehydrofrozen peas,' beef and pork raised on a new (hormone and antibiotic added) feeding method and lowfat milk."[43] Half a century later, if a president were to be fed such a miserable meal, he or she might feel entitled to send it back, with a rebuke to the chef. Eisenhower, however, seems to have been impressed.

The constantly heard refrain was that the American diet had never been better; and that it was the best in the world.[44] In 1952, the government's Food Protection Committee announced that "the American people now enjoy the most abundant and varied diet of any nation in history."[45] This was only possible, it added, because of improvements in food production and technology, above all the proliferation of chemical additives. At the beginning of the twentieth century, there were only about fifty additives in common use, and, like the benzoates in ketchup that Harvey Wiley had battled against, they had a murky image. After the Second World War, chemists came up with hundreds of new additives, which now had the air of a magician's box of tricks.

Like Eisenhower's dynamic conservatism, the postwar explosion of food additives represented a break with the past. There were countless new tools for extending shelf-life. New colourings gave processed foods a delightful illusion of freshness. A new breed of preservatives seemed to offer "virtual immortality for some kinds of baked goods."[46] As Ira Somers, an advocate for the food industry, explained, "in the United States a person buys a loaf of bread and it will keep for several days in the home without spoilage," whereas in those countries unfortunate enough to make their bread without additives, "there is considerable loss due to mould growth."[47] The obvious reply to this

229

"What will the world taste like tomorrow?" This advertisement from flavour manufacturer Norda International expresses the mood of promise in the flavour industry of the 1970s.

is that there are other, better ways of dealing with this problem; that buying good fresh bread on a near-daily basis, as is the French and German custom, or baking it yourself, can be a far preferable existence to having a permanent loaf of mould-inhibited "bread" in the bread bin. But the additives evangelists would have none of it. "Additives are needed to retain the standard of living to which we are all accustomed," claimed Somers. This was fast becoming the American Way.

You would expect an industry spokesperson to talk like this. What was more surprising was how ready government was to share the industry view. Despite the frequent complaints by the food industry that the FDA was obstructing its activities, government agencies under Eisenhower generally acted more to "allay public concern" over additives than to investigate whether such concern was justified.[48] The view of additives as both necessary and wonderful persists in the FDA to this day. On the FDA website, if you search for "additives," you will be directed to this series of questions, which have the bright tone of a 1950s' car salesman:

Q. What keeps bread mould-free and salad dressings from separating?

Q. What helps cake batters rise reliably during baking and keeps cured meats safe to eat?

Q. What improves the nutritional value of biscuits and pasta and gives gingerbread its distinctive flavour?

Q. What gives margarine its pleasing yellow colour and prevents salt from becoming lumpy in its shaker?

Q. What allows many foods to be available year-round, in great quantity and the best quality?

Answer: FOOD ADDITIVES[49]

Official optimism about additives, however, has always been met with widespread uncertainty about their safety. Looking beyond his bright patter, the 1950s' car salesman was selling a highly desirable product that could nevertheless kill its owner. Was the same true of the 1950s purveyors of substitute foods?

It was not long before this question came to the attention of lawmakers. In 1950, 1951, and 1952, Congressman James Delaney chaired a committee looking into the safety of chemicals in food. The committee reported that about 840 chemicals were currently in use in foods, of which only 420, or half, could be deemed "safe."[50] This worrying discovery led directly to the Food Additives Amendment of 1958, which was supposed to give the consumer greater protection against harmful additives. The new amendment was extremely complex, a reflection of the ambiguous status of additives. To start with, what *was* an additive, exactly? Weren't all ingredients additives, in a way? During the construction of the new law, one expert witness had suggested that cream might be considered as a "chemical additive" to ice cream—an obviously absurd notion, but one that could be overturned only by a very precise definition of the word "additive."[51] The new law defined additives as substances that leave residues in food or otherwise affect its characteristics—implying that it was distinct from the food itself. It then went on to put chemicals used in food in one of three categories.

First were those chemicals that must be excluded from food altogether—those that had been shown to cause cancer when ingested by either man or animals (the so-called Delaney Clause). Next came chemicals that must be subject to intensive testing by manufacturers and excluded from food by the FDA until proven safe. The third category of chemicals was very different. These were not prevented in any way from being added to the food supply, nor were they legally defined as "additives" because they had been "generally recognized as safe"—the so-called GRAS concept. A substance was placed on the GRAS list if it was deemed to have been "adequately shown through scientific procedures to be safe," or if it had been in common use in food prior to 1958. Some of these substances were traditional condiments such as salt, pepper, sugar, and vinegar.[52] Many more, however, were relatively modern chemical creations. The original list of GRAS chemicals numbered 182; by 1961, there were 718 chemicals on the GRAS list. If a manufacturer wanted to add a GRAS chemical to food, that was no one's business but the manufacturer's.

Thus, while the Food Additives law was designed to offer consumer protection, it offered a still greater protection to the manufacturer who wished to create "innovative" processed food. By the 1960s, the old recipe-based food standards of the Roosevelt years were starting to look creaky. In a world of powdered soup mixes, what use were recipes? In 1961, the FDA issued its first non-recipe-based food standard—for frozen raw breaded shrimp (prawns).[53] Instead of stipulating exact options for the ingredients of the batter and breading, it simply requested that "safe and suitable" ingredients should be used, at the manufacturer's discretion. This could include not just the ingredients most consumers would expect if they were breading their own shrimp at home—breadcrumbs, eggs, and milk—but "safe and suitable" emulsifiers, flavour enhancers, and preservatives. "Safe and suitable" was an all-encompassing category that could include countless substances that would once have counted as adulterants.

This new move was a sign that the FDA could no longer expect to maintain control over every new product that came into the marketplace. There were just too many of them, and most could not be measured by the old standards. By the 1970s, there were a thousand agricultural products in use in the United States, but twelve thousand natural and synthetic chemicals added either directly or indirectly during the processing of food.[54] Coming up with the standard for jam had been pretty easy, as we have seen; all the FDA had to do was to consult a range of traditional cookbooks and calculate a reasonable percentage of fruit to sugar. But no old recipes existed for most of the new- fangled processed foods being created. These confections existed in total opposition to home-cooked food. What cook would dream of making Razzles, Pop-Tarts, or Pringles (all junk foods launched in 1966–67)? Actually, there is a website now dedicated to people whose hobby is re-creating giant versions of junk food snacks (www.pimpthatsnack.com): crazed postmodern cooks spend days fashioning giant versions of Oreo cookies, or KitKats, or Jammy Dodgers. But the joke works only because cooking your own processed food is such an obviously perverse thing to do.

By the late 1960s, the old pejorative category of "imitation foods" no longer seemed appropriate for many of the processed foods around. It had been easy for Accum to complain about "lemonade" that was really water mixed with tartaric acid, when everyone knew it should have been made with fresh lemons. It was much harder to say what "Tab," the new diet drink introduced in 1963, was an imitation of. It might as well have appeared from space, so little resemblance did it have to any traditional beverage. Like so many of the new foods and drinks, it was a novel creation, *sui generis*.

In December 1969, President Richard Nixon called a White House conference on food and nutrition. The political route from Eisenhower to Nixon (via Kennedy and Johnson) marks a shift in national mood from prosperous optimism to cynicism and despair; and so it was with food. The backdrop to the conference was the discovery that many in America—especially the poor—were struggling with hunger and malnutrition. Far from being the best-nourished people in the developed world, as the Eisenhower-era zealots had proclaimed, Americans were among the worst. National statistics from 1967 showed that twenty-year-old men in thirty-six other countries would live longer on average than those in the United States. A study published in *Nutrition Education* in November 1969 revealed that nearly all American children under the age of one were deficient in iron. Meanwhile, obesity was on the rise. The study concluded that "Dietary habits of the American public have become worse, especially since 1960."[55]

The White House conference was primarily designed to address these problems. The government announced a huge expansion in the Food Stamp scheme and improved child nutrition programmes; better school lunches; and food education. Nixon launched a grand call to "put an end to hunger in America."[56] The conference also included a panel whose job it was to consider the question of "new foods." and what role they might play in saving the nation's diet. (No one seems to have thought it politic to point out that the decline in America's health had actually coincided with the proliferation of these foods.)

The New Foods Panel was chaired by the vice president of Monsanto (now most famous as a seller of GM technology, but then a producer of agricultural chemicals) and included among its members the vice president of Pillsbury (mass market baked goods) and the vice president of the Ralston-Purina Company (breakfast cereals and animal feed), as well as assorted food scientists and nutritionists. Unsurprisingly, given its industry component, the panel sagely concluded that novel foods were extremely valuable. It urged the complete modernization of food regulation, to permit "completely new foods" to be created.[57] It was noted by the panel that new foods were currently "often required by the Government regulatory agencies to be called 'imitation products,' *even when the new product is superior to the old.*" They went on: "The use of such over-simplified and in-accurate words is potentially misleading to consumers, and fails to inform the public about the actual characteristics and properties of the new product." Thus the official position on "imitation foods" had come full circle. The term "imitation" had originally been required to stop consumers from being misled. But now, the White House panel argued, the word itself had become misleading.

The New Foods Panel recommended a modernization of all food standards "to encourage the development and marketing of varia-tions of traditional foods, and of completely new foods, that can provide consumers a greater variety of acceptable, higher quality, and more nutritious food products at lower prices." The old recipe-based approach to food standards was criticized as "deadening." New Foods were the future, because they held out the alluring dream of solving the two great evils of the American diet: obesity and malnu-trition. They would do this with two powerful tools: the invention of novel slimming foods, and the fortification of staple foods. All that was needed was the creativity of the free market, coupled with a bit of government help when needed, and everyone would be better off. Or so it was hoped.

Fortification and Slimming

The idea of enrichment or fortification goes back to the early 1830s, when goitre, a grotesque swelling of the neck cause by enlarged thy-roid glands, was a common problem in many communities. This could lead in turn to cretinism, a form of mental illness. It was no-ticed that the places where goitre and cretinism were most com-mon were those where the soil was deficient in iodine. Add iodine to people's diets, and the goitre and cretinism never occurred. A French chemist, therefore, advocated adding iodine to all table salt. Routine enrichment of salt with iodine was introduced in Europe from the early 1900s; and, little by little, goitre became a forgotten disease in the affluent West, though in Pakistan, where not all table salt is io-dized, millions are still at risk from iodine deficiency.

This was an early, and isolated, case. More general fortification began only in the 1940s, as wartime governments panicked that the public was not getting enough nutrients from the food it ate. Knowedge of vitamins, those micronutrients so crucial to health, had been growing ever since 1897 when Christian Eijkman, a Dutch pathologist, discovered that eating unpolished rice could prevent beriberi because it contained thiamine. Each decade of the twentieth century brought a new vitamin or mineral to worship. In the 1900s, it was fish oil to cure rickets. In the 1920s, it was calcium and vitamin A, which led experts to recommend drinking enormous quantities of milk and stuffing oneself with green vegetables.[58] It also led to vitamin D being added to milk, to aid calcium absorption and prevent rickets. Next came vitamin C and vitamin G (later rechristened riboflavin). In 1940, in the United States, it was the turn of thiamine again, which became known as the "morale vitamin" in the fight against Hitler. Vice President Henry Wallace went so far as to say that adding thiamine and other B vitamins to the diet "makes life seem enormously worth living."[59]

The only reason they needed to be added to the food, however, was that the basic American diet was so depleted. Take bread. Most of the vitamins and minerals in flour are in the outer layers of the wheat—the bran.[60] The old method of milling white flour was to sift the crushed wheat through fine sieves or "bolting cloths"; this method retained many of the nutrients in the wheat. From the 1870s, however, a new, efficient system of roller-milling came in, which passed the flour between steel cylinders, stripping it of most of its vitamins on the way. Standard, "70% extraction" white flour will have lost 60 percent of its calcium, 77 percent of its thiamine, 76 percent of its iron, and 80 percent of its riboflavin. By 1940, the average American was eating 200 pounds a year of bread made from this nutrient-stripped flour. For the poor, this represented a very large percentage of their daily calories. Nutrition experts were terrified that America was suffering from "hungerless vitamin famine" as a result.[61] The government did not like to tell people to return to more nutrient-dense whole wheat bread; it wouldn't be popular. But during the 1930s, it had become commercially viable to produce vitamins on a large scale. So from 1940 onwards, inferior white flour was required to be enriched with thiamine, iron, and niacin.[62] This was followed in 1943 by enriched cornmeal and grits, and in 1946 by enriched pasta.[63] The same story unfolded in Britain, where flour was enriched with vitamin B and calcium from July 1940 on.[64]

Fortification had its sceptics. In Britain, Ernest Graham Little MP, a member of the Food Education Society, expressed doubts: "The universal scientific opinion is that the organic and natural supplies of vitamins are far superior to the synthetic kind."[65] In the United States, too, the American Medical Association cautioned against more extreme forms of "fortification." The temptation was for manufacturers to add nutrients to foods as a selling point, nutrients that often had never been there in the first place. SunnyFranks frankfurters boasted that they were enriched with vitamin D—"the Sunshine Vitamin hard-playing youngsters and hard-working men need! This Vitamin D does not 'cook out'!"[66] It was a pure marketing gimmick, an attempt to put these fatty sausages on a nutritional par with cod liver oil. Normal beef frankfurters are low in vitamin D compared to leaner red meats; dosing them up with the vitamin gave the impression that they were healthy, when actually a single frankfurter contains 20 percent of the recommended daily intake of fat. Other manufacturers engaged in a kind of "fortification race" to become the most highly vitaminized. Carnation bragged that its wheat contained "Actually 50% more Vitamin B1 than in whole wheat."[67] To innocent consumers, topping up on vitamins looked like an obvious benefit. How could you ever have too much of a good thing? In fact, though, it was just another swindle, and a potentially dangerous one.

In 1957, the Ministry of Health in Scotland reported adverse symptoms in a number of infants and young children: a failure to thrive, vomiting, weakness, and, in several cases, death. The problem was traced to excess levels of vitamin D in the food supply, because so many children were being fed up on vitamin D-enriched dried milk, plus cod liver oil. The children were suffering from hypervitaminosis D, an excess of vitamin D in the body, which leads to an overload of calcium, causing damage to bones, soft tissues, and kidneys.[68] The Ministry of Health intervened to reduce the levels of vitamin D consumed by children, and the problem abated.

Pillsbury Farina, a typical fortified food of the mid-twentieth century.

This was by no means the last case of vitamin poisoning, though. Illness from excess iron fortification has been on the rise for some time. As many as a million Americans suffer from a hereditary condition that means that they absorb slightly more iron than they need. On a normal diet, with not too much red meat, such people may be fine. If they consume too much iron-rich food, however, the iron can build up in their bodies to toxic levels, causing liver and heart problems and even death. From 1970 to 1994, iron in the American food supply increased by a third; in the same period, death by iron poisoning or haemochromatosis increased by 60 percent.[69] In 2004, the Danish government banned the sale of Kellogg's cereals on the grounds that consumers could overdose on the vitamins, especially pregnant women who could put their unborn babies at risk. This action was widely reported as "bizarre."[70] A spokesperson for Kellogg's announced that the company was "mystified." But the Danish decision was no more peculiar than the decision of earlier governments to endorse vitamin intake so wholeheartedly that consumers got the impression that it was impossible for added vitamins to do more harm than good.

Vitamins, like everything else, are a poison if taken to excess. In 2000, a study done by the U.S. Institute of Medicine reported that consuming large quantities of antioxidants (such as vitamins C and E and selenium) could lead to hair loss and internal bleeding.[71] The following year, three thousand children in Assam in India fell severely ill after receiving too high a dose of vitamin A.[72] Even when they are not toxic, high doses of single nutrients can create imbalances that affect the ability to metabolize other nutrients. They can also do damage by covering up the underlying flaws in the food to which they are added.

The New Foods Panel of 1969 had no reservations about what it saw as the advantages of food fortification. Its first recommendation was to urge "an immediate fortification programme to relieve malnutrition." The thought behind this was that too many staples eaten by poor consumers were simply "not nutrionally adequate." But the panel did not want to limit fortification to staple foods, insisting that "no one type of food should be preferred over another as a nutritional carrier."[73] This opened up the market for all manner of "enriched" foods, from doughnuts to candy. Sugary breakfast cereals could boast that they were good for bones or brains, or gave those who ate them amazing athleticism, because of their vitamin content. Instead of being a menace to public health, highly processed foods—if sufficiently fortified—could claim that they were improving it. In the trade press, the industry was a little more honest about who was really being enriched by enrichment. Hoffman-La Roche, a company selling bulk vitamins to manufacturers, urged an increase in voluntary fortification, arguing that "nutrition is good business . . . food fortification, technologically feasible at low costs, opens up new marketing possibilities for food manufacturers."[74]

Unlike iodine in salt, which addressed a problem affecting entire communities, this later wave of fortification attempted to address the health problems facing particular groups (the poor, children, the elderly, pregnant women) by dosing en masse the consumers of certain processed foods. This approach has its obvious drawbacks. One is that there is no guarantee that the fortification will reach those who most need it. Almost everyone eats table salt; the same is not true of expensive breakfast cereals. Another problem is that fortification may reach those who *don't* need it, who may actually suffer as a result. One example is folic acid in bread. Since the 1990s, it has been mandatory to add folic acid to bread and other grain products in the United States (and, at the time of writing, Britain looks set to follow, with folic acid added to sliced white bread). The reason is that folic acid consumed by pregnant women can prevent babies being born with neural tube defects, such as spina bifida. The downside is that universal fortification of popular foods with folic acid may harm older people, by masking a deficiency in vitamin B12. This deficiency affects up to 10 percent of those over sixty-five and can result in damage to the nervous system.[75]

There is another kind of deception going on too—a kind of collective self-deception. Fortification can disguise the fundamental inadequacies of the diet eaten by the general population. By bolstering the intake of certain select vitamins, fortification can give the impression that, in large industrial societies, the food of the poor or uneducated is not so much worse than the food of the rich or educated. This is an illusion. On grounds of both taste and nutrition, there is a great difference between eating a whole, tart, juicy orange, rich in fibre as well as natural flavour, and eating an orange-flavour drink fortified with vitamin C; or between eating a slice of real, malty wholegrain bread, naturally rich in B vitamins, and eating an industrially produced square of fortified white "bread." In this sense, fortification is a social panacea. As Marion Nestle writes:

> The fortification of cereals, milk and margarine . . . addresses vitamin and mineral deficiencies that are caused largely by poverty or other socioeconomic conditions that affect a relatively small proportion of the American population. In an ideal world, nutritional deficiencies among such groups would be corrected through education, jobs, or some form of income support—all better overall strategies than fortification.[76]

After 1969, fortification combined two contradictory views of the consumer. One saw consumers as children who need to be protected against nutritional harm without their knowledge; the other treated consumers as adults who can assume total responsibility for the food they buy. The basic premise of fortification programmes is that consumers are incapable of making sound judgements about what food would do them most good. If everyone ate iron-rich foods of their own accord, no one would suggest duping them into doing so. Fortification turns consumers into passive creatures, who swallow vitamins whether they choose to or not. Combined with this, however, was the idea of total consumer freedom. Nixon's New Foods Panel stated that "the consumer should be free to select, in the marketplace, any fortified food of her choice, whether completely natural or completely synthetic in origin."[77] This assumes that the consumer has an expert knowledge of nutrition. But if the consumer did have an expert knowledge of nutrition, why would fortified foods be necessary?

The New Foods Panel was at least confident that the risks of overdosage from enriched foods would not be anything to worry about. Maximum limits would be set on how much of any vitamin could be added to a given food, and the panel insisted that "overconsumption of any nutrient would be prevented by basing fortification on the calorie contribution of the food to the diet."[78] In other words, if you could work out how much bread or milk or cereal the average person ought to consume in the course of a day, you could make sure that he or she would not overdose on fortified foods. But there was a flaw in this reasoning too. Already in 1969, many Americans were not simply suffering from malnourishment but from overconsumption and obesity. It was all very well to set the amount of vitamins a given portion of food should contain. The trouble was that millions were not sticking to recommended portion sizes. If you binged on a whole box of vitamin-fortified cereal, you could very easily exceed the maximum limits on those "healthy" vitamins—never mind the damage to your health in other respects.

While much of the food industry was pressing hard to "nutrify" as many foods as possible, other manufacturers joined the drive to make new "nonnutritive" versions of traditional foods, in order to tackle the growing problem of obesity (and, of course, to turn a profit from it). Again, this development involved treating the consumer as someone simultaneously incapable of making choices and in need of an infinite array of them. In 1970, William F. Cody, a lawyer for a big food company, argued that, while it was true that "overweight can be largely eliminated through dietary adjustments," it was very difficult for the "average man" to change his dietary habits. The answer? "To provide palatable modifications of the high-calorie, high saturated fat foods; these special foods should look, smell and taste like the traditional food—but should be reformulated so as to reduce or eliminate the objectionable characteristics."[79] Cody gave the examples of low-calorie margarine and low-cholesterol, low-fat dried egg. He complained that current law still obliged these foods to be labelled as "imitation" foods, which "conjures up the image of something highly synthetic or cheapened." Cody pointed out that the old imitation foods were generally cheaper than those they were imitating. By contrast, these new "non-standard compositions" might actually cost more to manufacture, and therefore be more highly prized by the consumer. How could something be an imitaiton if it was the more valuable article? Low-calorie margarine was not the same as, say, watered-down milk, in Cody's view, because the low-calorie product was a deliberate improvement on the real thing: more expensive to make, better to eat, and healthier.

Unfortunately, whether it was healthier was becoming a moot point. In 1969, a crisis had hit for the fast-expanding diet industry when Robert Finch, the U.S. secretary of health, announced that he was ordering the removal of the artificial sweetener cyclamate

242

from the GRAS list.[80] Food scientists had been thrilled when they first discovered cyclamate in 1937. Like saccharine, it had sweetness without the calories of sugar, but unlike saccharine, it did not have a bitter aftertaste. In 1951, the FDA approved cyclamate, and it crept into countless slimming products, from chewing gum to diet soda, from children's vitamins to sugarless jam. When the Delaney Clause of 1958 prevented the FDA from clearing carcinogens as food additives, cyclamates—ironically—took over even more from saccharine, because of a 1951 report suggesting that the latter might cause cancer. By 1969, foods and drinks containing cyclamate were in as many as three quarters of American homes.[81]

Throughout the 1960s, however, studies had linked cyclamate with health problems. Studies in 1965 showed that cyclamates could cause cancer in lab rats, evidence initially discounted by the FDA. Further studies in Japan in 1966 showed that when cyclamates passed through the human body, they could create a different and highly dangerous chemical, cyclohexylamine (CHA). In America, several rats injected with cyclamates developed cancer of the bladder. Finally, in 1968, an FDA biochemist, Dr Jacqueline Verrett, did some experiments on the effects of cyclamates on chickens. She found that when cyclamate was injected into a hen's egg, the resulting chicks developed hideous deformities, "such as wings growing out of the wrong part of the body, a leg rotated in the socket, and extreme curvature of the spine."[82] These are not the sort of images you want in your head when giving your child a vitamin pill. Dr Verrett's results—which were publicized on TV, to widespread public horror—precipitated the nationwide ban on cyclamates in the United States and led to a voluntary withdrawal of cyclamates from many foods in Britain.

The cyclamates scandal did not, however, lead to a more sceptical attitude to diet foods and drinks in industry or government. On the contrary; with cyclamate gone, more slimming techno-foods were needed to take its place. There had been excitement in 1969, when the *Journal of the American Chemical Society* announced the accidental discovery of a new sweetener, made from amino acids and 180 times sweeter than sugar. It had been discovered by a scientist working at the laboratories of G. B. Searle, a big pharmaceuticals firm based in

This 1969 advertisement for Diet Pepsi boasts that
the drink is no longer sweetened with cyclamates.

Illinois. The scientist, James Schlatter, was trying to develop a treatment for ulcers. One of the protein compounds Schlatter created
spilled on his hand. Later in the day, he licked his finger and found
that it had a distinctly sweet taste.[83] By 1973, Searle was marketing
this sweetener as "aspartame," now famous as the essential ingredient
in Diet Coke. In 1974, the FDA cleared it for use in dry foods. Yet
again, however, the shadow of the Delaney Clause loomed.

In 1974, a neuroscientist, Dr. John Olney, came to the FDA with
reports of research he had done on aspartame in 1970, showing that
one of its components, aspartic acid, had caused brain lesions in mice
(something that Searle denied). Olney was joined by consumer lobbyists who protested that the approval of aspartame was premature.
The FDA froze its approval of the sweetener until more research could
be done. Finally, in 1980, a Public Board of Inquiry, which had been
convened to consider the safety of the sweetener, voted unanimously

to reject the use of aspartame—which now had the attractive name of NutraSweet—for the time being. The board took into account four studies. Three of them had concluded that aspartame was not carcinogenic; but one had reported an increase in brain tumours in rats. Taking a cautious view, the board concluded that there was a chance—however remote—that aspartame might be a carcinogen, and that it should not be marketed until further safety tests had been done.[84]

Just a year later, aspartame was once more approved by the FDA as safe for marketing. Why? The decision was influenced by economics and politics as well as science. By the end of the 1970s, there was enormous commercial pressure to clear aspartame for use. In 1977, the cyclamates nightmare was repeated when the FDA announced a ban on saccharin—then the only artificial sweetener permitted in the food supply—because Canadian tests showed that it could cause bladder tumours in lab animals.[85] The U.S. food market was shocked. Consumers stockpiled cans of their favourite Tab—the low-calorie drink sweetened with saccharin—before it was removed from stores.[86] Hundreds of workers at the Sweet'N Low plant in Brooklyn were put "on vacation." There was a frantic search for alternative sweeteners, to shore up the growing diet food industry.[87] The Department of Agriculture developed a sweetener called "narangin," made from citrus peel. The drawback: consumers registered its sweetness only after ten to twenty seconds of contact with the tongue, which was no good for dieters, who wanted instant gratification. Another sweetener, Miraculin, made from West African berries, seemed a winner until its manufacturers went bankrupt. This left only aspartame—whose sweetness was instant and which came with stable financial backing from Searle.

The problem was obviously the scientific uncertainty about its safety. Then again, a new study by the Ajimoto company, the Japanese manufacturers of aspartame, concluded that it was safe for human consumption, after all.[88] For the scientific community, this was just one more study—not the end of the story, but a springboard for further experiments. The science of risk can never eliminate uncertainty entirely. The law, however, cannot live in a state of flux. As one lawyer has written, apropos of the aspartame case, it is the job of the law to resolve questions "inexpensively and quickly."[89] In the end, it was politics that supplied the necessary confidence in NutraSweet.

Aspartame had been championed during the 1970s by Donald Rumsfeld, when he was chief executive of Searle, the company that manfactured it. In 1981, Rumsfeld moved to the White House, as Ronald Reagan's chief of staff. In that position, he oversaw the appointment of a new head for the FDA, who conveniently made the decision to overrule the board's initial decision and approve aspartame. The FDA now endorsed Searle's contention that there was no evidence of aspartame having a carcinogenic effect on the brains of animals; the science and statistics on which the fourth report had been based were flawed. In 1983, Coca-Cola announced the distribution of Diet Coke sweetened with NutraSweet, and before long it was one of the world's most ubiquitous forms of sweetness. By 2005, it was being used in more than six thousand food and drink products around the world, including Diet Snapple, Sugar-Free Kool Aid, low-calorie crisps, chewing gum, yoghurts, vitamins, medicines, and desserts.[90] Pepsi and Coke may be rivals in many respects, but they are united in their use of NutraSweet.

Meanwhile, consumer groups, spearheaded by campaigning attorney James Turner, have continued to raise doubts about the safety of aspartame.[91] In 2005, it was widely reported that an Italian research project based at the Ramazzini Foundation in Bologna involving eighteen hundred lab rats had found an increase in lymphomas and leukaemia among the female rats who were fed aspartame. Spokespeople for aspartame, however, vigorously disputed the research results, insisting that they was "inconsistent with human epidemiological data." The International Sweeteners Association commented that "with billions of man-years of safe use, there is no indication of an association between aspartame and cancer in humans."[92]

Over the past twenty-five years, a great deal of research has concluded that aspartame is safe for human consumption.[93] Much of this research has been funded by the aspartame industry, but some has been independent. As recently as 2006, a huge federally funded study on human beings, rather than rats, conducted by the National Cancer Institute in the United States found no increased cancer risk even among people who drank vast quantities of Diet Coke and other aspartame-sweetened sodas.[94] Enemies of aspartame claim that the components of aspartame—especially aspartic acid and phenylalanine—can be associated with adverse symptoms, such as headaches and seizures. Aspartame's advocates reply that "These components are consumed in much greater amounts from common foods, such as milk, meats, fruits and vegetables."[95] Nevertheless, in 2007, a major British supermarket, Sainsbury's, announced that it was removing aspartame from all of its own-label soft drinks in response to customer concerns, and replacing it with sucralose, yet another nonnutritive sweetener.[96]

Who is right? Who can say? I would not dream of impugning the integrity of either Donald Rumsfeld, Diet Coke, Nutra Sweet, or the FDA. But nor am I keen on consuming aspartame on a regular basis, if I can avoid it; not because I think it is bad for me, but because I dislike the taste and find the notion of all noncalorific sweeteners—whether sucralose, saccharine, aspartame, or cyclamate—mildly creepy. The food writer Amanda Hesser has written of her prejudice against artificial sweeteners on the grounds that "I avoid eating food whose ingredients I can't picture in my head." She remarks that "When you taste raw sugar, you can make out the grains on your tongue, and as they melt, your palate gets a gentle, glorious bath of sweetness and caramel," whereas when you taste a sweetener "it gives you a jolt of sweetness and zero nuance."[97]

Safety aside, the most interesting aspect of the explosion in slimming foods in the postwar years was the way that ersatz foods effortlessly transformed themselves into consumer desirables. They violated one of the most fundamental properties of food—which is to nourish—but did so with such swagger that you might think they were saving the human race. Scientists boasted of creating "nonfoods" that could be eaten "with pleasure and interest" without making anyone fat: "attractive meals of guaranteed physiological valuelessness."[98] In 1968, the General Foods Corporation took out a patent for making artificial fruits and vegetables, made from "an edible, crisp, chewable non-uniform agglomerate of calcium alginate cells." These would be crunchy, like vegetables, but almost entirely lacking in food value. Another nonfood was the synthetic cherry, made by dropping cherry-coloured sodium alginate solution into a bath of calcium salt; over time, little cherrylike droplets would gel together. These non-cherry cherries had the "advantage" that they were not affected by the heat of an oven. By 1970 synthetic cherries had been success-fully marketed in the United States, Australia, Holland, France, Italy, Switzerland, and Finland.

How was such a thing possible? How did these nonfoods man-age to please consumers (or at least some of them), when the er-satz foods of the early twentieth century had caused such fear and loathing? The answer lay not just in the very different circumstances in which they were produced: war and peace. They also owed their success to the machinations of the food technicians who laboured to make nonfoods "acceptable" to the average palette. Their chief tool in this quest was an astonishing new array of flavourings.

The Brave New World of Flavourings

In the early 1970s, a firm named Stevenson & Howell ran an adver-tising campaign in the food trade press. The ads featured pictures of mutant fruits—a pineapple crossed with an apple, or a plum that had mated with a strawberry, or a weird "pear-nana," half banana, half pear. The tag-line read: "We can create any flavour you like. Try us."[99] It went on: "There's nothing our research boffins enjoy more than to be given a big, juicy problem. One that can really stretch them. If the flavour you have in mind doesn't exist at the moment. Fine. If it does exist but you want an improvement on Nature. Even Better." An ad for a rival firm, Florasynth, showed a blond-haired toddler sucking his thumb. "There are very few flavours we can't reproduce" was the slogan.[100]

During these halcyon years for synthetic flavourings, there was a mood of excitement and possibility among the scientists who chose the career of the commercial "flavourist." Gone was the furtive shame of the flavour adjusters of the nineteenth century, the crude fraudsters exposed by Accum and Hassall who used chilli to spice up ginger or citric acid in the place of lemon. These new flavourists

Do you feel like a pearnana?

It happens. You're eating away and suddenly oranges, apples and pears just taste plain ordinary.

There's no surprise any more.

You want something different. But what? A pearnana, for instance?

We've grown big feeling that way at Stevenson & Howell. Because our researchers are forever searching for new flavours. And, more often than not, come up with them. Flavours that can be marketed in dozens of different products.

So why not set us a challenge?

Think exotic. Pull out a flavour that's been skulking in the back of your mind and let's see what we can make of it.

Creating flavours is food and drink to us.

Stevenson & Howell   S&H

Stevenson & Howell Ltd. Standard Works, Southwark St. London SE1.
Tel: 01-928 4833. Grams: Distiller, London SE1.

222   The Flavour Industry MAY 1972

The "pearnana"—a 1970s' trade advertisement from Stevenson & Howell.

saw themselves as artists, operating with the blessing of the government, Willy Wonkas who could fashion entire meals out of nothing. They had an elevated view of their calling: "It is by the artistry of the flavourist that a flavour stands or falls," said one.[101] In exercising

249

their highly attuned senses to harness a whole spectrum of flavours, they were like the haut-parfumiers of Paris, except that instead of essence of violets or tincture of rose, they used tomato extenders and powders of artificial cheese. The similarity was no accident. The flavourings industry emerged initially as a subsection of the perfume industry. Both depended, first, on the science of essential oils, and later, on "synthetics." The world's biggest flavour manufacturers—IFF in the United States and Givaudan in Switzerland—still create perfumes as well. As the Givaudan website says, it is all part of the same "sensory innovation," whether the thing being created is body lotion or a beef stock cube.[102]

Like the perfume artists of France, the flavourists used the vocabulary of music to describe what they did: the harmony of the perfect flavour; the scale of sweet, sour, salty, bitter.[103] In his 1970 book, *Synthetic Food*, Magnus Pyke compared the incredible new tastes coming out of the flavour labs to the inventiveness of great composers: "After all, a Beethoven symphony is an artificial mixture of noises unknown to nature, but, in many people's opinion, better than any natural sound. The taste of Coca-Cola is also unknown to nature."[104] In 1973, one flavour chemist wrote that the true flavourist "must have virtuosity . . . as does a musician in command of his instrument: a complete confidence in his own ability to extract from his medium the full scope of possible effects."[105] It sounds so impressive, you could almost forget what the virtuosity was for, namely. creating fake prawn cocktail flavouring for crisps or black cherry flavour for frozen dessert.

The new music of flavour came in many different forms. At the more traditional end, there were single-note essences such as lime, hazelnut, or coffee. Also common were "oleoresins," concentrated solvent extractions from certain natural ingredients, such as paprika, cloves, black pepper, celery. More exciting to the flavourists were "seaslics" or liquid savoury flavourings, which represented "total flavourings" for such products as pâté de foie, liver sausage, or Polish sausage. Powdered flavour mixes gave the plethora of new snack foods their moreish qualities. Then there was the whole family of flavour extenders and enhancers, which tricked the palette into thinking that what it was tasting was realer than it was. With "tomato extenders," manufacturers could drastically cut down on the amount of real tomatoes they needed to use. As with the adulteration of the past, so it was with these new flavours. One additive led to another. Flavourings often needed "flavour adjuncts" before they could be added to food. Extenders—such as cellulose—could "carry" or "extend" a feeble flavour to make it seem richer and longer in the mouth. Spray-dried emulsified oils gave "mouthfeel" to soups and sauces, deluding you into thinking that there was more to them than powder and water. Flavour enhancers—such as MSG (monosodium glutamate)—and flavour suppressors—such as sucrose—could mask unpleasant flavours and magnify desirable ones. More recently, a substance called Miraculin has been developed that changes sour into sweet in the mouth.[106]

A waste of good science? The flavourists didn't think so. They took pride in the thought that they were creating an entirely new sensory landscape. In the prewar period, commercial food flavouring was not so very different from what the cook might do at home with a collection of spices, zests, and essences. In 1922, a Mr. A. Clarke, a technical food adviser to various British firms, revealed that a standard commercial sarsparilla flavouring consisted of sarsparilla root, sassafras bark, and licorice extract—a very natural combination. A spicy "cake flavour" mix might be nothing more than cinnamon bark oil mingled with clove oil, bitter almond oil, lemon oil, and sweet orange oil—a delicious-sounding mixture entirely in keeping with the traditions of home baking, the only difference being that the home cook would probably use orange and lemon zest, powdered cinnamon and clove, and almond essence.[107]

Admittedly, some artificial flavours were already in use. Fruit-flavoured confectionery was often made from a mixture of natural fruit essences with artificial fruit "esters." These are compounds made by the reaction of an acid and an alcohol. Following their discovery in the mid-nineteenth century, esters were used in the manufacture of synthetic eau de colognes and floral extracts. For example, *Geranyl Propionate* was found to have an odour resembling bergamot, while *Styrolyl Valerianate* had a strong scent of jasmine. Other esters smelt fruity and were used to boost natural fruit essences. *Ethyl Cinnamate* had a nose uncannily like ripe apricot, while *Phenyl-Ethyl Acetate* was peachy. *Isopentyl acetate* has a powerful odour with a distant resemblance to pears. It would come to be used in the tooth-cracking boiled sweets called pear drops. Many English children were fond of them. No one could pretend, however, that pear drops really taste like pears. Their flavour is one-dimensional, basic. Continuing the musical analogy, pear drops were like a symphony with all the instruments missing except for one.

By contrast, the postwar flavourist's art was all about complexity and creating flavours that could almost stand in for the real thing. Thanks to the advent of gas chromatography, a technology that could separate out many different chemicals from a complex compound, the flavourist enjoyed a hugely widened palette. In 1966, a "tolerable" artificial flavouring for pineapple consisted of no fewer than ten chemical compounds mixed with seven natural oils.[108] Now, the figure would be much higher. In *Fast Food Nation* in 2001, Eric Schlosser listed the flavour ingredients that go into a typical artificial strawberry flavour, "like the kind found in a Burger King strawberry milk shake":

> Amyl acetate, amyl butyrate, amyl valerate, anethol, anisyl formate, benzyl acetate, benzyl isobutyrate, butyric acid, cinnamyl isobutyrate, cinnamyl valerate, cognac essential oil, diacetyl, dipropyl ketone, ethyl acetate, ethyl amyl ketone, ethyl butyrate, ethyl cinnamate, ethyl heptanaote, ethyl heptylate, ethyl lactate, ethyl methylphenylglycidate, ethyl nitrate, ethyl propionate, ethyl valerate, heliotropin, hydroxyphrenyl-2-butanone (10 per cent solution in alcohol), α-ionone, isobutyl anthranilate, isobutyl butyrate, lemon essential oil, maltol, 4-methylacetophenone, methyl anthranilate, methyl benzoate, methyl cinnamate, methyl heptine carbonate, methyl naphthyl ketone, methyl salicylate, mint essential oil, neroli essential oil, nerolin, neryl isobutyrate, orris butter, phenethyl alcohol, rose, rum ether, γ-undecalactone, vanillin and solvent.

To a cook, this list of ingredients is disturbing and alien. In the spring of 2004, a panel of food experts discussed junk food at the Manoir aux Quat'Saisons restaurant in Oxfordshire. Alice Waters was there, the chef of Chez Panisse in California. She mentioned with horror the passage about the strawberry milkshake in Schlosser's book. Didn't everyone know, Waters asked, that the ingredients of a strawberry milkshake were simply strawberries, milk, sugar, and a little ice cream?

Flavourists, not being cooks, would see this as the wrong way of posing the question. They would say: of course you can make a delicious strawberry milkshake if you have the finest organic strawberries at your disposal, the creamiest milk, the best cane sugar. To a flavourist, the challenge is to make a convincing strawberry milkshake using no strawberries or milk, and perhaps no sugar either. To aid them in this task, they need as many chemical tools as possible. It was partly the overwhelming number of different flavour components that gave the flavourists of the 1960s and 1970s such a sense of freedom. Their "library" of ingredients eventually ran into thousands. By 1986, it was calculated that between 3,500 and 6,000 flavours were used in British food, "depending on who is doing the counting."[109] Flavourings accounted for over 95 percent of all additives used.[110]

Moreover, the use of these flavours, in contrast to the use of preservatives, colours, and other additives, was hardly policed. In 1980, a former Unilever scientist boasted that in Britain "flavours are unique among food additives insofar as they have defeated all efforts to bring them under satisfactory statutory control."[111] He added, almost as an afterthought: "not that the public is thereby exposed to unnecessary hazard." In the United States, there have been lists available of those flavourings "generally recognized as safe"; the first list contained approximately eleven hundred ingredients, but it has mushroomed since then.[112] Until 1999, there was no European inventory of permitted flavourings.[113] Anything went, and for the most part still does. Even now, there is no requirement, either in Europe or in the United States, for flavour companies to disclose the components of their flavours on food labels. Their presence is advertised only under the blanket term "flavour" or "artificial flavour."

For pure-food campaigners, this enormous and seemingly unpoliced growth in flavourings was a dystopian horror, but to the flavourists, the infinite permutations of artificial flavours offered a bright new dawn in which flavours that were once the preserve of the rich were now within everyone's reach. "What will the world taste like tomorrow?" asked big flavour firm Norda International, adding that its creation of new flavours meant that "there will be more to taste in the world tomorrow."[114] "Felton makes the world taste better" was the slogan for another international flavour company, which advertised itself with pictures of happy-looking, bright-eyed multiracial children, all enjoying the sheer goodness of synthetic flavouring in its various forms: a lurid soda pop, an artificial cake, a breakfast cereal, a strange green ice cream cone, and a giant swirly lollipop.[115] The implication was clear: gone was the austerity of the postwar years; gone was the dreariness of boiled cabbage and corned beef. Thanks to the wonder of flavour technology, no one need stint themselves or their children any more. It was Space Invader crisps and cheap ice lollies all around. A flavour scientist named R. H. Sabine claimed that the new flavours were part of "improvements in standards of living." The poor, who could not afford fresh fruit, could at least enjoy "the flavour of fresh fruit."[116] Never mind that this flavour lacked any of the nutrients of the real thing—who would deny those smiley children their share in the wonderful world of taste-sensation?

From today's perspective, when manufacturers are generally keen to hide their white-coated food technicians away, to preserve the illusion that their products are all "natural," it is startling to look back on this more shameless era, when food science was openly proud of its role in the future. Flavourists would talk brazenly of their role in undermining "natural" food. They also knew what their gold standard was. An industry textbook writes of the invention of cola flavour as "the dream of all flavourists: to create a new, non-existing taste with a world-wide success which by now lasts 100 years." In flavourist's terms, cola is the perfect flavour, carefully balanced between citrus top notes, a "sweet spicy, cinnamon, creamy vanilla heart," and an earthy sourness underneath. "The cola flavour is an unequalled success story since the invention of cough syrup by Dr. Pemperton."

In much of the trade advertising for the flavour industry of the 1970s, you find an open contempt for the food of nature—so messy, so expensive, and so hopelessly unreliable compared to the miracles

Felton
makes the world
taste better.

The right flavor can make the difference   *Felton*   Felton Co. (G.B.) Ltd., European Head Office
between cornering a new market and sharing an   Pondwood Close, Moulton Pk., Northampton NN3 1RT
old one. A Felton flavor makes a world of difference. Manufacturing facilities & branches around the world.

Australia · Canada · France · Great Britain · Hong Kong · Italy · Japan · Mexico · Spain · United States · Venezuela

"Felton makes the world taste better." This optimistic 1970s'
advertisement stands in stark contrast to today's worries
about the effects of additives on children.

in the flavourist's test tubes. "Durkee improves on the fickle tomato,"
boasted one advertisement.[117] Synthetic flavours, by contrast, held
out the promise of "relatively stable prices, freedom from quality
variation and regular availability."[118] "In memory of our dear beloved
raspberry," reads a 1975 promotion for White Stevenson flavours. A
real raspberry is shown dead, in a glass coffin. "But why worry," it
cheerily continues. "The raspberry still lives on in Reigate's new Essence and Natura. With remarkable accuracy these products capture
the flavour that was the raspberry."[119] Flavourists appear to have had

255

something of a fixation with raspberries. Another 1970s advertisement, this time for Barnett & Foster, shows a white-coated scientist using a compass to measure a giant plastic model of a raspberry, as if trying to reduce its elusive flavour to an incorruptible formula.

In real life, raspberries all taste different. That is their joy. A truly sweet raspberry, bursting with juice without being overripe, is all the more lovely for the knowledge of other, more disappointing raspberries: mouldy ones, or sour ones, or hard, mean, seedy ones. But flavourists don't see it like that. A current textbook on flavourings complains that "common cultivated raspberries" often have a taste that is "just watery and acidic." Raspberry flavouring, on the other hand, is the distillation of the most fragrant ripe rasberries ever tasted, with "a delicious fresh, fruity, green, floral, violet like perfume with some seedy woody background."[120] Broken down, the essential code for raspberry is as follows. For the floral violet perfume, mix in some alpha- and beta-ionone. For the fruity raspberry body, a touch of 1-(4-hydroxyphenyl) butan 2-one. For the fresh green top notes, a soupçon of (Z)-3-hexenal. If a jammy quality is desired, it can be supplied by the merest hint of 2,5-dimethyl- 4 hydroxy-furan-3(2H).

This is only a basic blueprint. The real trick was in the flavourist's individual tweaking, which all came down to his particular aesthetic judgement. The Holy Grail was finding the precise raspberry flavour mix that would hook consumers and make them come back to *your* raspberry-flavour dessert rather than those of your competitors. The flavourist's ultimate job was "to evoke pleasurable recognition in the mind of the taster,"[121] because this recognition could make the difference between profit and loss for the firm he worked for (and in these advertisements and manuals, the flavourists are always a he; women, like children, exist to be seduced by the new flavours). Like the haut-parfumier second-guessing the secret desires of his exclusive clientele, the top flavourist had to predict exactly what "raspberry" meant for possible consumers, some of whom may never have tasted a real one. He must know his market. For example, in the Netherlands, consumers like their dried packet chicken soup mixes to come flavoured with just a suggestion of curry. This wouldn't work in England, where dried chicken soup, curiously, needs a base note of sage, perhaps a remnant of the old tradition of sage and onion stuffing.[122] If asked about it, most consumers wouldn't know that the sage was there; but they would miss it if it was absent.

Things didn't always turn out quite so well. In 1970, Tudor crisps announced triumphantly that it had succeeded in creating the perfect kipper-flavoured crisps. For years, fish-flavoured crisps had been an impossible dream, perennially thwarted by technical difficulties. Fresh fish was just one of those flavours that didn't translate into flavour powders. At last, in 1969, flavourists working for Tudor had a eureka moment: why not substitute kippers with their strong, salty, smoky taste, for fresh fish? After months of research and numerous taste panels involving adults and children, the perfect kipper crisp was finally formulated, using a combination of "fish concentrate and special smoked flavour." A packet was designed showing a bright red fish. The crisps went on sale on 22 June in Scotland and 29 June in the north of England. Tudor's marketing manager explained that he hoped they would be a hit among "kipper connoisseurs."[123] Foolish man. Just because the flavour of kipper crisps was technically accurate did not mean people wanted to eat them. Kipper lovers are generally traditional folk, who want their kipper flavour in a real kipper, not in a packet of crisps. Kipper crisps developed a modest following—and are still sometimes remembered by nostalgic children of the 1970s—but they were ultimately a failure, compared to the enduring British crisp flavours of smoky bacon, salt and vinegar, and cheese and onion.

A successful flavour was one that created brand loyalty. Lucas Seasonings boasted that it had "the sort of natural taste that raises a product above the rest and makes customers come back for more of the same."[124] Another firm called Spice & Flavour offered "tailor-made" seasonings, rubs, sauce mixes, brine flavourings, flavoured pie gelatines, pastry glazes, smoke flavours, breadings, batters, and specialized meat tenderizers. It argued that using its services "could mean the difference between Brand leader and also-ran." Its ads played on the idea that consumers may buy a packaged product expecting a certain taste that it then fails to deliver. "He's just been sold a pack of lies," says one. We see a picture of a sad-looking old man holding up a package of what looks like chicken stew. Clearly it is not to his liking. "Flavouring is a subtle business. Often it's the label that dictates the taste; tells you the taste you should expect, but don't get. And that's not right. Or good enough." At this point, Spice & Flavour positions itself as the saviour. "Call in the favour detectors from Spice & Flavour Services. They're experts. They analyse taste. Follow it. Anticipate it. Capture it." The striking thing is the implication that Spice & Flavour is engaged in the science of detection, when its real business, clearly, is the science of deception. The business of adulteration has been turned on its head. All the howls of indignation, all the pleas for moral integrity and claims of scientific accuracy—the tools in trade of the nineteenth-century anti-adulteration campaigners—have been stolen by the adulterators themselves.

How did the flavourists justify their part in this? There were essentially three defences: pleasure, price, and chemistry. Pleasure, because as flavourists like to say, "the flavour and fragrance industry is solely aimed at enhancing the human striving for increased pleasure and sensual enjoyment. . . . Hedonistic aspects, therefore, form the basis of our industry."[125] If you should reply that the hedonistic enjoyment of food long predates synthetic chemicals, or point out that the enjoyment to be had in a synthetic soup mix, for example, is a pretty poor kind of pleasure, then the flavourists will hurl the question of price in your face. The science of flavour makes food much cheaper than it otherwise would be. The flavourists present their work as being essentially philanthropic, an argument that might be stronger were it not for the fact that the economic savings are more to the benefit of the manufacturer than the consumer. As one advertisement from 1975, directed at producers, not consumers, advised, use Ottens artificial bacon flavour and you'll "laugh all the way to the bank."[126] Anyway, it's all so unnatural, you reply; so horribly *chemical*. Now the flavourist plays his trump card. *All* flavours are chemical, he says. Consumers are wrong to mistrust "chemicals" in their food when all food is made up of chemicals anyway. In the words of one industry expert: "What is it that makes one food product insipid and almost flavourless and another exceptionally flavourful and appetizing? Let us not be in any doubt. In all cases, it is the presence, or absence, of chemicals."[127] So there.

We need to remember, these are chemists talking. Chemically speaking, there is no difference between the natural flavours in natural food and chemical re-creations of those components. Take vanillin. In 1873, the German scientist W. Haarmann found that it was possible to isolate vanillin, one of the most highly flavoured components in vanilla: $C8HbO_3$ (4-Hydroxy-3-methoxy-benzaldehyde). Once isolated, this chemical could also be reproduced. Another German scientist, Karl Reimer, found that vanillin could be synthesized from guaiacol, a kind of yellowish, aromatic oil derived from creosote. Today, thousands of tonnes of vanillin are made, usually from byproducts of sulphite waste.[128] For the flavourists, it would be wrong to be too squeamish about this. Vanillin is vanillin is vanillin, whether it comes from curving pods on a beautiful, orchidlike plant growing on the island of Madagascar, or from industrial waste. Its effects will be the same.

The law agrees. In Britain, flavours such as vanillin that occur naturally in food are called "nature-identical." The label does not have to state where they actually come from. A flavouring counts as fully "artificial" only if it does not occur in nature at all, as is the case with another, stronger vanilla-substitute called ethyl-vanillin (often used in chocolate). Vanillin is now the most commonly produced flavour compound in the world, with an annual market of approximately 12,000 tonnes.[129] It is not hard to see why, when you discover that real vanilla extract costs (at 2004 prices) around seventy-three U.S. cents per gallon whereas extract made from synthesized vanillin is around twelve cents. As a result, almost all cheap "vanilla–flavour" cookies and ice creams and cakes are actually flavoured with vanillin.

Does this matter? Can we accept the flavourist's defence? To answer this, let's consult the law on flavouring. Under European Union law, there are two basic rules governing whether flavourings should be allowed in food. The first is that "they present no risk to the health of the consumer."[130] On this count, vanillin is probably fine. There are no records of anyone ever getting vanillin poisoning, no matter how many fake vanillin cookies they consumed. As of 1980, the UK guidelines were for a maximum of 20,000 mg of vanillin to be added per kilo of food.[131] This is extremely high, a sign of vanillin's low toxicity. Compare this figure with the guidelines for piperine, one of the components of black pepper: a maximum of just 1 mg per kilo. It would be hard to make the case that vanillin poses health risks.

But what of the second EU rule? This states that the use of flavourings "must not mislead consumers." The flavourists would exempt themselves from blame here, too, saying that so long as the label does not pretend that a food contains real vanilla, there is no deception in the use of vanillin. But can we be so sure? Vanilla expert Tim Ecott writes that the prevalence of vanillin is such that "many consumers have never tasted the difference" between vanillin and vanilla.[132] To a real vanilla lover, there is a world of difference between vanillin, with its one-dimensional sweet, creamy taste, and a real vanilla pod, whose sweetness is tempered with a warm, winey woodiness. Vanillin is merely one of several hundred chemicals in vanilla. Yet to those who have never tasted anything else, manufactured vanillin seems like the one true taste. Ecott writes that "Tests conducted by flavour manufacturers have revealed that many people prefer artificial vanilla, simply because it is the taste they know."[133] The flavourists have succeeded in convincing the world's taste buds that "vanillin" is actually "vanilla."

Ultimately, the brave new world of flavouring is founded on deception. The better the flavours, the worse the deception. Human beings have nine thousand taste buds, which send us complex evolutionary signals about which foods are safe and good to eat. In manipulating these taste buds, the flavourist convinces the public that food is something other than what it really is. What makes it even worse is that the particular foods on which the flavourist has always practised the craft are the ones most lacking in nutritional values, or the ones with the most to hide. There were even glimmers that the flavourists knew this was what they were doing. In 1972, the flavour industry announced an exciting new development called Interchicken, a method for injecting "flavour compositions" into "intensively reared poultry," to make up for the blandness of these poor unhappy creatures. A trade report admitted that "it is well known that poultry slaughtered for processing after only 49 days is somewhat deficient in flavour compared with a bird held for a period of 112 to 120 days." The answer? Inject the birds with flavours using "an electronic dispensing unit," which dispersed flavour throughout the flesh. The flavour mix included "an autolysate produced from virgin yeast, a concentrate derived from New Zealand butter known as Butta Natura and chicken spice liquid based on extracts from a wide range of herbs and spices, all dispersed in soluble grade phosphate and citrate."[134] Interchicken went down well with test panels; it convinced them that a modern broiler could taste as good as a properly reared chicken.

As Eric Schlosser has said, the essential job of the flavourist is "to conjure illusions about processed food."[135] But back in the 1970s, these conjurors were not having it all their own way. While the flavourists were happily composing their ersatz music, campaigners both in Britain and the United States had begun to rebel against the manipulation.

Ralph Nader and the Chemical Feast

In 1973, a witty food industry executive coined a new term: "Naderphobia," named after the attorney and consumer activist Ralph Nader. The symptoms of the illness were a heightened sensitivity and an abandonment of normal behaviour among otherwise happy-go-lucky businesspeople; in other words, a terror in the face of the damage that consumer activists could do to the food industry.[136] Plenty of manufacturers were suffering from this malaise. Ralph Nader's main target, famously, was the automobile industry. From 1965 onwards, Nader had attacked General Motors for producing cars that were "unsafe at any speed." Since then, his advocacy of consumer causes had widened out to include drug companies, air pollution, and foods that were unsafe in any amount. In 1970, Nader's Study Group published *The Chemical Feast*, a blistering attack on the American food supply and the FDA's failure to police it.

Nader directly compared his work to that of Upton Sinclair. "We're Still in the Jungle" was the title of an essay from 1967. Like Sinclair, he ruthlessly exposed the fact that the meatpacking industry remained—sixty years on—inadequately policed, leading to gross malpractice. History was repeating itself:

> About 15 percent of the commercially slaughtered animals (19 million head) and 25 percent of commercially processed meat products in the U.S.—enough meat for 30 million people a year—are not covered by adequate inspection laws. According to the Department of Agriculture, significant portions of this meat are diseased and are processed in grossly unsanitary conditions, and its true condition is masked by the latest preservatives, additives and colouring agents.[137]

That is to say, the same, sorry old swindles went on, despite decades of federal food law and despite the apparent plenty of postwar America. Hams were still pumped with water to boost the weight. Tainted and rancid meat was still offered for sale—in one year, twenty-two million pounds of it were seized. Old meat was still freshened up to look as good as new. Nader exposed "hamburger embalmed with sulfite, a federally banned additive that gives old meat a deceptively bright pink colour." He claimed that a New York state survey found sulfite in twenty-six out of thirty hamburger samples.[138] (In Britain, a similar practice went on without breaking the law. Processed meats could be dosed with Ronoxan D20, an antioxidant that prevented "undesirable effects such as white spot in sausages." A trade advertisement for Ronoxan showed a butcher's display of raw sausages, boasting, "They'll look as fresh tomorrow as they did today.")

It was the same old story, but worse. Nader argued that advances in food technology enabled swindling to take place on a far greater scale than in Sinclair's day. At least the foul practices of Packingtown had been limited by the crude tools available. It "took some doing to cover up meat from tubercular cows, lump-jawed steers and scabby pigs." Now, however, "the wonders of chemistry and quick-freezing techniques provide the cosmetics of camouflaging the product and deceiving the eyes, the nostrils and taste buds of the consumer. It takes specialists to detect the deception. What is more, these chemicals themselves introduce new and complicated hazards unheard of 60 years ago."[139]

Together with his groups of eager student helpers, who became known as Nader's Raiders, Nader railed against the "chemical feast" of the average American diet. He took on everything from unclean poultry plants to fatty hotdogs.[140] Fighting bad food, for Nader, was a mission central to democracy. "What we need is a much healthier citizenship if the government is going to be preserved," he announced.[141] Pulling the wool over consumers' eyes was a fundamentally undemocratic activity. Nader wanted to empower ordinary shoppers to take back control of what they ate.

In his crusade against bad food, Nader did not cut an entirely sympathetic figure. Like Hassall and Wiley before him, he sometimes let his single-minded fight for purity get out of hand. Charles McCarry, a biographer who followed him around in the early 1970s, recorded his behaviour towards those who served him impure foods:

> Nader is not a gallant customer. With stewardesses and waitresses he is relentless: to offer him a soft drink is to invite a detailed analysis of the harm done to the human body by a drink filled with sugar and caffeine. The suspicion he feels for American food transfers to those who serve it; he glowers at the sore-footed women in restaurants and at the jaunty miniskirted girls in the aisles of jet airplanes as if they are, all of them, unwitting Borgias. "The only thing you should be proud to serve on this whole airplane," he says to one puzzled stewardess, "is the little bags of nuts. And you should take the salt off the nuts."[142]

Nader's justification of this churlish behaviour would be that, in complaining so relentlessly about the evils of modern food, he was benefiting everyone, even the poor saps who had to serve it.

It was largely thanks to the lobbying of Nader's Health Research Group that MSG (another substance that had been innocently sitting on the GRAS list) was removed from baby foods. Nader's Raiders also inspired other liberals to set up consumer groups. The lawyer Dr. John Banzhaf had his own group of lobbyists, nicknamed Banzhaf's Bandits, law students at George Washington University in Washington, DC, who fought against the deception of food advertising. In its posters, the Campbell Soup Company had concealed marbles in the photographed bowls of soup to make the liquid appear thicker and creamier than it really was. Banzhaf's Bandits succeeded in getting the company to discontinue this practice, though it could not persuade Campbell to print a correction, apologizing for having used the marbles in the past.[143]

Another campaigner against the lies of the food industry was Dr. Robert Choate, a wealthy civil engineer with a startling physical resemblance to Abraham Lincoln. Choate financed his own campaign against hunger and "nutritional suicide." In 1970, he told a Senate subcommittee that many of the best-selling American breakfast cereals were virtually worthless from a nutritional standpoint.[144] Choate complained that these cereals were "huckstered" to children with commercials on Saturday morning television. An example was Kellogg's Sugar Frosted Flakes, which was sold with the slogan "Eat . . . and you'll be a tiger in no time." Thanks to Choate's efforts, the majority of cereal recipes were improved somewhat.[145]

These small consumer triumphs were not to be discounted. Too often, however, lobbyists would campaign against one swindle or poison, only for another to spring up. In February 1976, under pressure from Nader's Health Research Group, the FDA banned Red Dye no. 2, or amaranth (E123),from the food supply. Until that point, Red Dye no. 2 had been the most widely used of all food colourings, present in "almost every processed food we eat," lending its purple-red glow to ketchups and condiments, candies and jellos, sausages and chocolate cake.[146] Some 4.5 million dollars' worth was produced every year, colouring 10 billion dollars' worth of food: this despite the fact that doubts had been cast on the safety of Red Dye no. 2 for twenty years. It was first named as a suspected carcinogen at a conference in Rome in 1956. It was banned in the USSR in the 1960s after Soviet scientists linked it to cancer. In a good piece of Cold War politics, the FDA dismissed this Soviet study as "inconclusive," only to backtrack in 1976 and declare that it was a potential carcinogen, after all. The ban on Red Dye no. 2 is often cited as one of Nader's great successes in the realm of food.

Whether the ban reduced the chemical load of the average consumer is unlikely. After the ban came into effect in 1976, there was a huge increase in the production of another red food dye, Red no. 40 (Allura Red or E129). Nearly two million pounds of Red no. 40 were certified for use that year. Though more orangey in colour, like Red

Dye no. 2, Red no. 40 could be used in almost every processed food, from the coating for fried chicken to lurid red soft drinks. Yet, in the opinion of some, it was actually more harmful than Red Dye no. 2. Like Red Dye no. 2, Red no. 40 is an "azo dye," made from double-bonded nitrogen. In December 1976, a consumer group headed by Michael Jacobsen argued that Red no. 40 "should never have been admitted into our food supply," citing studies involving mice who developed malignant lymphomas after eating the dye.[147] Jacobsen pointed out that Red no. 40 was banned in many European countries. (Today it is still banned in Denmark, Belgium, France, Germany, Switzerland, Sweden, Austria, and Norway.) His words fell on deaf ears. Red no. 40 was too valuable to the food industry. Consumer confidence had already been shaken by the Red Dye no. 2 affair, which had led to many mothers picking the red ones out of packets of multicoloured candies. Where would it all end? The federal government chose not to act against Red no. 40.

As a result, Red no. 40 is still present in countless foods in America. It has been largely cleared of the charge of being a carcinogen, but it is known to produce skin allergies in some people. Several scientific studies have also confirmed that consuming Red no. 40 has a bad effect on the behaviour of children suffering from attention deficit disorder.[148] More informally, many parents have given accounts of the almost immediate tantrums that can ensue in some children after just one portion of food coloured with Red no. 40.[149]

The affair of Red Dye no. 2 versus Red no. 40 showed the problem of fighting additives in the food supply one by one. This approach was all very well in the days of Accum, when the chemicals of deception were much fewer in number, and much more obvious in the ways they did harm. By the late 1970s, though, the danger was everywhere. At that time, a California expert on allergies, Dr. Ben Feingold, noticed some startling results when "difficult" children who had hitherto been drugged with tranquillizers and sedatives were able to function normally without drugs on a diet that completely eliminated all colourings, flavourings, and salicylates. This became known as the Feingold Diet. Its approach of near-total abstinence from additives addressed the fact that in real life, additives are seldom eaten one at a time. Safety tests for additives were generally carried out on isolated substances. In practice, though, the average consumer of processed food would eat an unpredictable cocktail of chemicals, which might react with each other in unforeseen ways.[150] In 1985, it was calculated that the average British Christmas dinner might contain as many as 170 different additives, all in a single meal.[151] Some campaigners in both the United States and Britain were starting to see that the real battle was not so much between this or that additive, as between a diet laced with ersatz everything and a life of eating real food.

Caroline Walker and Legalized Consumer Fraud

In the early 1980s, a young nutritionist named Caroline Walker tried to awaken the British public to the poisonous tricks of the food industry. Walker, who, from surviving photographs, had a wide, open smile, was educated at the genteel Cheltenham Ladies College (where she took A levels in biology, chemistry, and art). From an early age, though, Walker shunned gentility to become a political radical. She soon decided that the biggest scandal in Britain was the falsified state of the nation's food. She called it, straightforwardly enough, *The Food Scandal* in her book of 1984 (coauthored with Geoffrey Cannon), which became a number one bestseller. Having trained as a nutritionist, Walker took every opportunity she could to lecture, write, broadcast, and complain about the "counterfeit" state of British food. "Lemonade contains no lemon," she opined, "cheese and tomato snack no cheese. In 1978 23,000 tonnes of cheese went into manufacturing. In 1983, the quantity was almost halved to 13,700 tonnes. Why? Cheese analogues, that's why."[152] It made her sick.

Like George Orwell, Walker saw ersatz food everywhere; but even Orwell never imagined anything quite as "unfoodlike" as the additive-laden junk exposed by Walker. "Consumers are beginning to learn that what they're eating isn't food, but chemistry sets," she said, only half joking.[153] In the 1970s and 1980s, British food regulation had followed the American pattern. Gone was the old emphasis on recipe-based food standards, with the exception of a few "traditional" foods such as butter and corned beef. In 1973, Britain had entered the European Economic Community, which meant more trade with its European neighbours. In this new commercial paradise, too many food standards could be a drag; they would slow down the wheels of commerce. The Food Standards Committee advised that so long as "foods are labelled, advertised and promoted for what they are, as required by law, there can be no objection to their sale."[154] Walker complained that decisions such as these were generally made—in an undemocratic state of secrecy—by "middle-class, middle-aged men" in Whitehall, "who don't cook and who don't go shopping."[155] Many of these men, in her view, had no idea how bad the food in the shops really was.

Thus far, the story of swindling has been mainly a story of men; of the men who swindled and of the men who tried to expose them. The bulk of those being swindled, however, were women, since they were generally the ones who bought the household's food. Most of the food advertising of the 1960s and 1970s was aimed at "Mrs. Average Housewife," a figure "whose only kitchen tools are a pair of scissors for cutting polythene and foil wrappings, and a tin opener." As Derek Cooper remarked in 1967, advertisers worked hard to convince Mrs. Average Housewife that she "is far too busy compulsively forcing grey out and white in to her husband's shirts to spend more than a few minutes a day at the cooker."[156] The entire edifice of postwar processed food depended on seducing Mrs. Average Housewife into thinking that ease and cheapness mattered more than quality. A trade advertisement from 1975 showed a beautiful blonde woman with a buttercup under her chin. The tagline was "She'll like margarine too if it's coloured with betacarotene."[157] The thinking behind such marketing was that Mrs. Average Housewife was so emptyheaded you could convince her of just about anything.

With acid wit, Caroline Walker cut through these deceptions. Like Mrs. Average Housewife, she knew her way round a supermarket aisle. Unlike her, she was not seduced. One of Walker's most powerful strategies, when giving a public lecture, was simply to bring along sackfuls of processed food and then reveal the ingredients. Often, at the end of a lecture, a man—one of those "middleclass middle-aged men"—would come up to Walker and say, "It's fascinating. Where did you get this stuff ?" to which she would reply, "Where do you think I got it? In the shops!"[158] In 1986, she was speaking at a conference on chemical additives at the Dorchester Hotel in London. Her audience consisted of industry representatives, nutritionists, and journalists. She dramatically produced from her bag a "disgusting drink," a laser-beam blue concoction called Mixed Fruit Tropic Ora, and asked if anyone would care to try a sip. Not a single food industry rep took up the challenge, despite the fact that they were responsible for producing this drink, or others like it. Finally, she "embarrassed" the food writer Paul Levy into tasting the blue liquid, which he pronounced "the nastiest thing I have ever tasted."[159]

Walker had a strong sense of history. She often voiced her admiration for the work of Frederick Accum and Arthur Hill Hassall. Like Accum, she saw adulteration as a consequence of industrialization, coupled with an excessively laissez-faire political class. She saw an important difference, though, between their campaigns in the nineteenth century and hers in the twentieth. Accum and Hassall had exposed *illegal* food fraud, whereas Walker's mission was to criticize swindles against consumers that broke no laws: she called it "legalized consumer fraud." It was perfectly legal, for example, to sell a "Raspberry Flavour Trifle" that contained no hint of real raspberries (under the absurd legal sophistry that if it had been called "raspberry flavoured trifle" rather than "raspberry flavour trifle" it would have been required to have *some* raspberries, while "raspberry trifle" would have had more raspberries still). Walker listed the ingredients in one such raspberryless trifle:

Raspberry flavour jelly crystals: sugar, gelling agents (E140, E407, E340, potassium chloride), adipic acid, acidity regulator (E366), flavourings, stabilizer (E466), artificial sweetener (sodium saccharin), colour (E123)

Raspberry flavour custard powder: starch, salt, flavourings, colours (E124, E122)

Sponge: with preservative (E202), colours (E102, E110)

Decorations: with colours (E119, E132, E123, E127)

Trifle topping mix: hydrogenated vegetable oil, whey powder, sugar, emulsifiers (E477, E322), modified starch, lactose, caseinate, stabilizer (E466), flavourings, colours (E102, E110, E160 (a)), antioxidant (E320)

Disgusting, I hear you say. I never eat those things. Well, someone does.[160]

Walker knew that the Ministry of Agriculture was happy to defend such aberrations on the grounds that "food is part and parcel of the enjoyment of life," and that the occasional fake raspberry trifle never did anyone any harm. Walker disagreed. The point about such monstrosities was that they rapidly came to be seen as "normal" food. Manufacturers preferred them to real food because they were so much cheaper. Soon, she noted, "the raspberryless trifle becomes the rule rather than the exception."[161] Once that happens, it becomes harder and harder for a more scrupulous manufacturer to make a real raspberry trifle, from real raspberries, real custard, and real sponge. The standards for raspberry trifle become debased. Trifle itself becomes reduced, from an occasional splendid treat to an everyday occurrence, no more special than packet blancmange or packet jelly, yet another daily injection of chemicals, sugar, and fat to add to the mix swirling in your stomach.

Walker, who was nobody's fool, was fond of remarking that whenever people like her—middle-class women with an axe to grind— criticized the additives industry, they would be met with two replies from arrogant food scientists: "potatoes and preservatives."

At conferences on food safety and quality, you know what you're in for when the lecture on additives starts like this. Food scientist strides confidently on to platform. Demands first slide. Up comes picture of potato, whereby lecturer delivers brisk commentary on poisonous chemical—solanine—naturally occurring in humble spud (potatoes are a real favourite for natural poisons). Nervous titters from the audience. Second slide: picture of a miserable slave in Ancient Egypt dipping meat into vat of preserving salts.[162]

269

The potato argument was that since some poisons occurred in nature, food has never been totally safe, therefore modern additives were nothing new. "Preposterous," said Walker. "If food does contain natural poisons, that is no reason for adding more." The preservatives argument was that "additives" have been used ever since ancient times, and, without them "we would all be keeling over with food poisoning." Another piece of specious logic, in Walker's view, since "arguments in support of preservatives are of no relevance to any other additive." In the mid-1980s, preservatives actually accounted for less than one percent of all additives used. "All the rest are flavours, colours and processing aids, which are of greater benefit by far to the manufacturers than to the consumers.... Why should manufacturers be allowed to hoodwink the public in this way?"[163]

Caroline Walker hoped to effect a major political change in what was done to police this deception. Sadly, she never got the chance. She died in 1988, aged just thirty-eight. She had been suffering from colon cancer. The irony was not lost on her. She told Derek Cooper, presenter of the BBC *Food Programme*, that the doctors couldn't believe it at first, thinking, "Here's this young woman, a nutritionist of all things, stuffing herself with wholemeal bread and she goes and gets her guts ruined!" She continued in her conviction that most cancers were diet-related, though. She speculated that hers had been caused by her seven years at Cheltenham Ladies College, where there was a "lack of fresh food, lack of whole food—sticky buns, sweets, white bread and margarine."[164]

If she failed to demolish the system of 'legalized consumer fraud," Walker did at least leave behind some advice on the best way that individuals could protect themselves against it. The "Caroline Walker Nutritional Guidelines" are still generally reckoned to be the best measure for testing the nutritional content of school meals in Britain, and they underpinned Jamie Oliver's recent battle to improve school food. Like Accum, 160 years earlier, Walker saw that the best weapon against adulteration was a knowledge and enjoyment of good food. Like him, she saw "good quality wholemeal bread" as the cornerstone of a happy life. She advised people to avoid processed fake sugars and replace them with the sugar in real fruit; to avoid fake hydrogenated fats and replace them with real sunflower, olive, and walnut oils; to shun soft drinks and colas and drink real juice instead; to eat fewer highly processed salty foods, but more flavourful herbs and spices; to eat mackerel and sardines instead of fatty meats and pies. Her entire message was crystallized in four simple words: "EAT WHOLE, FRESH FOOD."[165]

6

BASMATI RICE AND BABY MILK

We have traded away quality for a false sense of safety.
—Real Milk Campaign (2006)

Adulteration, mercifully, is a thing of the past. Such, at any rate, is the consensus in much of the secondary literature on the subject. Modern societies should be grateful not to have the problems of those wretched Victorians, with their watered-down milk and their alum bread. An academic research paper on Victorian adulteration concludes that we should be "very glad" not to be eating Victorian food.[1] A British educational website about food for secondary school children informs us that "In the past, people found lots of sneaky ways to make extra money, by adding dodgy ingredients to food and then making them look like the real thing." Now, though, we have food laws to "protect your health and stop you being cheated."[2] Thank goodness for that.

Much of this is wishful thinking. Food fraud is no more a relic of history than greed and deceit are. "Adulteration is hardly mentioned today," commented a historian of spice in 1993.[3] Maybe so, but only because talk of "adulteration"—such an unwieldy word, too close to adultery—has been replaced with endless discussion of food piracy, food scams, food fakes, food fiddles. As we saw in chapter 1, food swindlers thrive whenever there is a long chain between consumer and producers; now that chain is longer than ever, with food being freighted across the globe and back by vast anonymous distributors.

Every week seems to bring a new headline reminding us that our food is only as trustworthy as the people who make it and sell it. "Hazard reported in apple chemicals";[4] "Chilean grape scare";[5] "Cooking oil scandal hurts Spain";[6] "Pet food sold for human use";[7] "More food contaminated with dye";[8] "Indian clampdown on adulterated milk";[9] "Free-range scandal dupes consumers."[10] Food scares have become so ubiquitous, we no longer know what to make of them. The alarm is sounded; the offending food is withdrawn; those guilty of selling it apologize; and then we all go about our business again until the next food scandal breaks.

Food swindling is a more diversified business than it was in the past. It assumes many more different forms. Fear pulses around food like a constant background noise. This fear can lead in turn to a paranoia about food that opens up new markets in swindling, ones that prey on people's anxieties. These are the omnipresent tricks of the "legalized consumer fraud" exposed by Caroline Walker, which have taken a disturbing new turn with the falsification of agriculture itself, as intensive modern farming changes the basic nutrient properties of foods. But there are still plenty of illegal swindles going on too. Because the science is so sophisticated now, these can appear complex and hi-tech compared to the deceptions of the past. In the case of Basmati rice (which we will come to), the fraud can be detected only at the minute level of DNA. In essence, though, the crime itself still consists of nothing more than substituting a less valuable rice for a valuable one. There is still a market for the age-old tricks of watering down, colouring up, or bulking out the expensive with the cheap, as well as straight substitution and mislabelling. This goes on far more than is generally acknowledged, often in the highest reaches of the food business.

It wasn't meant to be like this. Everything was supposed to be different, in this era of "transparency" and "traceability." In the 1990s, with the birth of the "Knowledge Age," Western governments put their faith in the power of the label finally to solve all the problems in the food supply. Those alive and eating food today have more information about it at their disposal than Accum, Hassall, and Wiley would ever have thought possible. Whether their diets have been saved by it is another matter. It turns out that not all information about food is equally useful.

The Perfect Label

In 1990, Dr. David Kessler, a young and clever man with degrees both in law and medicine, became the latest in a long line of FDA commissioners, stretching back to Harvey Wiley in 1906, who came in to office determined to sort out food fraud once and for all. Kessler was sworn to the post on the same day that the new Nutrition Labelling and Education Act was signed. He made it his mission to protect consumers from misleading information about food—and he was going to do it through labelling.

In July 1991, *Time* magazine applauded the optimistic Kessler for his "utterly novel vision that consumers should be able to tell what they are ingesting by reading what is written on food labels."[11] It was a bit much to call it "novel" when Harvey Wiley had had the same idea almost a hundred years earlier. On the other hand, Kessler, unlike Wiley, had succeeded in putting his idea into practice, despite energetic counterlobbying from the food industry. Thanks to Kessler, 1993 saw a radical new system of food labelling in the United States, which required all food packages to list "Nutrition Facts," in a form that consumers could understand. Kessler called it a "public health opportunity of enormous magnitude," that "Americans will be able to pick up a food and see that one serving contains, for example, twenty-eight per cent of a daily intake of fat, or sodium, or fibre or cholesterol."[12] For Kessler, this was a "revolutionary" development.

He had a point. By 1990, food labelling—in the United States as in Europe—was in chaos. A third of packaged goods had no nutrient information at all; a third, mainly fortified foods, were *required* to list nutrient content; and a further third included some nutritional information voluntarily. Dr. Louis Sullivan, secretary for the Department of Health and Human Services, described the situation before 1993 as a "Tower of Babel," with consumers needing to be "linguists, scientists and mind readers to understand the many labels they see."[13] By contrast, the new labels of 1993 presented the content of food in terms

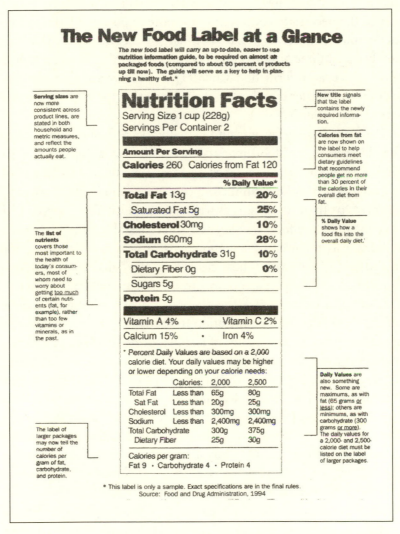

The New Food Label at a Glance

The new food label will carry an up-to-date, easier to use nutrition information guide, to be required on almost all packaged foods (compared to about 60 percent of products up till now). The guide will serve as a key to help in planning a healthy diet.*

Serving sizes are now more consistent across product lines, are stated in both household and metric measures, and reflect the amounts people actually eat.

New title signals that the label contains the newly required information.

Calories from fat are now shown on the label to help consumers meet dietary guidelines that recommend people get no more than 30 percent of the calories in their overall diet from fat.

% Daily Value shows how a food fits into the overall daily diet.*

The **list of nutrients** covers those most important to the health of today's consumers, most of whom need to worry about getting *too much* of certain nutrients (fat, for example), rather than too few vitamins or minerals, as in the past.

The label of larger packages may now tell the number of calories per gram of fat, carbohydrate, and protein.

Daily Values are also something new. Some are maximums, as with fat (65 grams *or less*); others are minimums, as with carbohydrate (300 grams *or more*). The daily values for a 2,000- and 2,500-calorie diet must be listed on the label of larger packages.

Nutrition Facts

Serving Size 1 cup (228g)
Servings Per Container 2

Amount Per Serving

Calories 260   Calories from Fat 120

	% Daily Value*
Total Fat 13g	20%
Saturated Fat 5g	25%
Cholesterol 30mg	10%
Sodium 660mg	28%
Total Carbohydrate 31g	10%
Dietary Fiber 0g	0%
Sugars 5g	
Protein 5g	

Vitamin A 4%	•	Vitamin C 2%
Calcium 15%	•	Iron 4%

* Percent Daily Values are based on a 2,000 calorie diet. Your daily values may be higher or lower depending on your calorie needs:

	Calories:	2,000	2,500
Total Fat	Less than	65g	80g
Sat Fat	Less than	20g	25g
Cholesterol	Less than	300mg	300mg
Sodium	Less than	2,400mg	2,400mg
Total Carbohydrate		300g	375g
Dietary Fiber		25g	30g

Calories per gram:
Fat 9 · Carbohydrate 4 · Protein 4

* This label is only a sample. Exact specifications are in the final rules.
Source: Food and Drug Administration, 1994

"Nutrition Facts": David Kessler's radical overhaul of the American food labelling system in 1993 was supposed to protect the consumer.

that regular folks could relate to: calories, calories from fat, cholesterol, sodium, total carbohydrate, etc. Another triumph was the fact that Kessler's FDA managed to secure more or less universal compliance from manufacturers, even though most had complained about what they considered an abominable restriction. Kessler announced

275

that by December 1994, more than 99 percent of packaged goods carried the necessary labelling.[14] Even more importantly, what information there was seems to have been largely accurate. The FDA randomly tested three hundred products and found that 87 percent had entirely accurate information, while 93 percent gave accurate statistics for calories. Kessler told the *New York Times* that the new system deserved "an A-plus."

The new labelling law was undoubtedly a victory of openness over the old secrecy that food manufacturers had got away with for so long. A scientific article on labelling from 1993 refers to "the age of the enlightened and discriminating consumer" and looks forward to an era of "truth in advertising as well as truth in labelling," which had been the dream of food campaigners since Hassall's time.[15] The 1993 U.S. law was followed by a comprehensive new food labelling law in Britain in 1996, with robust clauses against "misleading descriptions" and new requirements to display nutritional information.[16] This was a welcome contrast to the 1980s, when an atmosphere of official secrecy prevailed in relation to food, as if the ingredients in a Cornish pasty were a matter of state security. In 1998, an amendment improved the British law still further by adding requirements for Quantitative Ingredient Declarations (QUIDs). These meant that if someone sold a steak and kidney pie, for example, they needed to state how much steak and how much kidney was in it; if a pie contained only measly amounts of steak, the customer could see for themselves and choose not to buy it (though in 1998, the last year of the BSE crisis in Britain, many consumers preferred to eat pies that contained no steak at all).

Labelling became the great panacea of food safety during the 1990s, and it seemed there wasn't a problem it couldn't solve. Labels could signal to coeliacs that a sauce contained traces of gluten, tell vegans that a pudding had beef gelatine in it, and warn allergy sufferers that a bagel had been made in a factory that also used nuts. The age of information would answer everything. This has continued in the twenty-first century, with the new "traffic light system" of labelling in Britain, where consumers can see at a glance through a system of different-coloured stickers whether a food is healthy or not.

So why have the results in terms of food quality and consumer awareness been so underwhelming? There is no question that we have far more information about the food we eat available to us than any previous generation. Compared to Arthur Hassall in the 1850s, who had to put food under the microscope to find out its ingredients, we are immeasurably lucky. Yet labelling has not done away with swindling. Much of what passes for information in food is really an assemblage of meaningless statistics. Information assumes an audience capable of being informed by it; and the indications are that most consumers are not. For all its virtues, a reliance on food labelling to combat adulteration has at least four major drawbacks.

Supplement Facts
Serving Size 1 Packet

Amount Per Packet		% Daily Value	Amount Per Packet		% Daily Value
Vitamin A (from cod liver oil)	5,000 IU	100%	Zinc (as zinc oxide)	15 mg	100%
Vitamin C (as ascorbic acid)	250 mg	417%	Selenium (as sodium selenate)	25 mcg	36%
Vitamin D (as ergocalciferol)	400 IU	100%	Copper (as cupric oxide)	1 mg	50%
Vitamin E (as d-alpha tocopherol)	150 IU	500%	Manganese (as manganese sulfate)	5 mg	250%
Thiamin (as thiamin mononitrate)	75 mg	5000%	Chromium (as chromium chloride)	50 mcg	42%
Riboflavin	75 mg	4412%	Molybdenum (as sodium molybdate)	50 mcg	67%
Niacin (as niacinamide)	75 mg	375%	Potassium (as potassium chloride)	10 mg	< 1%
Vitamin B$_6$ (as pyridoxine hydrochloride)	75 mg	3750%			
Folic Acid	400 mcg	100%	Choline (as choline chloride)	100 mg	*
Vitamin B$_{12}$ (as cyanocobalamin)	100 mcg	1667%	Betaine (as betaine hydrochloride)	25 mg	*
Biotin	100 mcg	33%	Glutamic Acid (as L-glutamic acid)	25 mg	*
Pantothenic Acid (as calcium pentothenate)	75 mg	750%	Inositol (as inositol monophosphate)	75 mg	*
Calcium (from oystershell)	100 mg	10%	para-Aminobenzoic acid	30 mg	*
Iron (as ferrous fumarate)	10 mg	56%	Deoxyribonucleic acid	50 mg	*
Iodine (from kelp)	150 mcg	100%	Boron	500 mcg	*
Magnesium (as magnesium oxide)	60 mg	15%			
			* Daily Value not established		

Other ingredients: Cellulose, stearic acid and silica.

"Supplement Facts"—rigorous labelling was also brought in for vitamins and mineral supplements.

First, while the labelling regulations of the 1990s put a stop to many of the old deceptions of concealment in food, they gave encouragement to a new kind of deception, the technically true statement that can nonetheless mislead. Now that they had been forced by law to include so much data on their labels, manufacturers fought back by adding some more of their own. "Low in fat" read the label on countless boxes of cereal. This was no lie—this was no chicory labelled as "coffee." Yet there was something not quite right about it. For what cereal *is* high in fat? There has been a similar trickery in much of the labelling that adorns the food we eat: "reduced salt," boasts a packet of crisps, without indicating, needles to say, that it is still high in salt compared to other foods. Other packets sell themselves with little logos of hearts or brains or pregnant bellies, to indicate some wonderful medicinal property, which is invariably far from unique to the food in question. "Disease-specific" marketing of food is closely policed by the authorities, but the big food producers have their marketing people constantly thinking up ways around it; ways to use the label as a tool of advertising rather than education.

This proliferation of printed words and symbols on the side of food has led to a second drawback. Such is the information overload now that many consumers are more often baffled than enlightened by labels. As a senior official at the Food Standards Agency told me: "There's a problem that a lot of consumers don't understand the information they are getting on the label. They don't like numbers and percentages."[17] The more baffled consumers are, the less they will bother to read labels at all, never mind act on them. Many will simply give up. From 1993 to the present day, during the same period that Kessler's shiny new nutrition labels have been in effect telling the public what portions to eat, obesity in the United States has risen from 23 percent of the population to more than 30 percent. At its worst, the information age has meant reinforcing people's capacity to be cheated, rather than their right to good food.

The third drawback of labels in the information age is that they can lead us to overlook all the information that is *not* on the label. Some of this information remains secret thanks to legal loopholes. As we will see later in this chapter, a bread label in Britain must include all the "ingredients" but is not required to list processing aids such as enzymes, despite mounting evidence that these leave lasting traces. An apple pie may state the fact that it contains apples, but not whether those apples contained pesticide residues. A can of tomatoes may say "packed in Italy," without admitting that the true country of origin was China. A tin of "pitted black olives in brine" may not fully explain on the label that the only reason the olives inside are black is that they have undergone an oxidation process that colours them prematurely; in every other respect, they are essentially green olives. The vast majority of "black" olives sold are of this kind; yet they are still allowed to call themselves "black olives" on the label.

Finally, an obsession with labelling can lead us to forget how much food is altogether unlabelled. This includes the most honest food, as well as some of the most dishonest. You don't need a label to tell you of the excellence of the lamb chops you buy from a reputable butcher; or the quiveringly ripe melon you pick out from your favourite market stall in June; or the tangy unpasteurized Cheddar you choose from an old-fashioned dairy. You need only your senses, and a knowledge of the person you are buying from. This is a much more meaningful kind of knowledge than any label can supply.

On the other hand, some unlabelled food is more dubious. Much food still evades the stranglehold of the label altogether because it is sold in unpackaged form by restaurants, takeaways, cafés, and kiosks. A label is only as good as the packet it comes attached to. We are encouraged to think of our dependency on packaged food as a bad thing. But whenever we buy anything unpackaged, we are just as dependent as we ever were on the integrity of the people who serve us.

The horror stories are not hard to find. In September 2006, a German meat distributor committed suicide after police impounded 120 tonnes of "putrid" meat, which had been intended for sale as doner kebabs at stalls in Germany. A Munich police officer spoke of a "doner kebab mafia" engineering the fraud, which involved recycling meat well past its sell-by date.[18] In 2004, in the so-called Surrey Curry scandal, public analysts found that there were "illegal and potentially dangerous" levels of food colouring in chicken tikka masala purchased from curry houses around Britain.[19] (To which the *Sun* newspaper responded, "Don't Nikka Our Tikka," as if it were a terrible piece of busybodying to try to make curries more wholesome.) But it is not just cut in cut-price fast-food joints that such things happen. Routine deception occurs even at some of the world's most expensive restaurants.

Gourmet Food Fiddles and Protected Foods

In chapter 3, we saw how the poor of Victorian London had an incentive for colluding with the swindlers who defrauded them, because adulterated food was often the only kind they could afford to buy. The rich have different incentives for keeping quiet about those who swindle them, mostly the desire not to look stupid. Mark Leatham is a quality-food merchant, the founder of Britain's Merchant Gourmet brand, which sells such treats as Puy lentils, garlic-infused oil. and *dulce de leche* caramel; he also supplies many restaurants. "This business is full of hoodwinking," he remarked when I interviewed him in 2006; other food merchants I have spoken to have confirmed this view. Leatham told me of numerous fiddles in the quality food market, which usually go unnoticed, either because people don't like to complain or because they don't know the difference between one food and another. Leatham mentioned farmers in Yorkshire who send their sheep off for a "two-week holiday" in Wales; they come back tagged as "Welsh lamb" and appear as such on restaurant menus.

Leatham's most dramatic example concerned caviar. For a while in the early 1990s, Leatham supplied caviar to various grand establishments. It was a short-lived endeavour, he says, "due to the fraudulent nature of the suppliers (who would sometimes supply the wrong type of roe) and their up-market customers (hotels and casinos whose chefs were invariably on sizeable back-handers)." While he was still involved in the caviar business, Leatham tried to persuade one of the fanciest hotels in London to take his caviar; let's call it "The Glitz."

Leatham brought along Beluga (the most expensive caviar, prized for its large sweet roe), Oscietra (the next expensive, whose eggs are smaller and whose colour varies from dark green to golden), and Sevruga (a grey and nutty caviar, at that time the least expensive, though the world caviar shortage has since pushed prices up). The chef of the Glitz immediately said, "Oh we don't use Oscietra," in the tone of voice that implied that Oscietra was not quite good enough for his brand of clientele, though actually most gourmets regard it as the most delicious. "We only take Beluga and Sevruga." (The real reason for this, says Leatham, is that posh hotels liked to offer Beluga for clients who wanted to pay the most, and Sevruga for those who wanted to pay the least; taste hardly came into it.) Leatham and the chef began to taste the different caviars. First, they tried the Sevruga, tasting Leatham's Sevruga against the one that the Glitz currently used. Both were good. Next, Leatham persuaded the chef to try just a spoonful of the Oscietra. He agreed it was fabulous but regretted that he had no use for it. Finally, they came to the Beluga. The chef opened up his prized tin of Beluga. "Look, that is what you have just eaten, it's Oscietra!" exclaimed Leatham. The chef, deeply embarrassed, and inwardly furious at his supplier, had to agree that it was. He had been swindled, and was swindling his customers in turn.

Do such swindles matter? There are certainly worse things than fleecing a little extra money out of the very rich, who buy Beluga only as a status symbol anyway, caring little for the taste as they flush it down their gullets. But the swindlers are doing more than just puncturing the pretension of those with more money than sense. They are attacking the integrity of the food itself, which makes more widespread swindling easier to achieve. They are presuming—often rightly—on the ignorance of the consumer; and in so doing, they are perpetuating that ignorance.

Such is undoubtedly the case with another luxury item, saffron. Saffron adulteration is one of the oldest tricks in the book, because the real thing is so labour-intensive to produce. To make a pound of spice, you need as many as 200,000 little saffron threads, stigmas from a certain breed of crocus. Having always been prized and expensive, it has always been falsified. In the fourteenth century, in Nuremberg, saffron adulteration became so widespread that the city passed a special saffron law, the *Safranschou*, governing the inspection of saffron. Those who broke this law were thrown into the *Loch*, the deepest hole at the bottom of the dungeon in Nuremberg gaol.[20] Yet still the deceptions went on. Merchants would mix in a few orangey-gold marigold petals, or soak the saffron in honey to increase the weight.

From what we can tell, saffron swindles are just as common today. Many is the tourist who has bought a wonderfully cheap packet of "saffron" on their travels to Marrakesh or Istanbul, only to bring it home and find that it is a mixture of turmeric and food colouring. There is an easy test you can do to verify if your saffron is genuine. You need no scientific skill. Drop a pinch of it into a glass of warm water. If it takes a few minutes for the colour to diffuse, darkly, out of the strands, then it is the real thing. But if it colours the water yellow straight away, you know it is fake and you have been duped. Sadly, by the time you do the test, it is probably too late, because you are a plane-ride away from the person who sold it to you.

More strikingly, saffron swindling still goes on in grand restaurants, even though you are face to face with the swindler and could potentially confront him or her. Mark Leatham tells me a story about taking his saffron producer out to lunch one day at one of the poshest Moroccan restaurants in London. They ordered two saffron dishes. Neither of them came with saffron; both were the bright crude colour of turmeric. They sent the food back to the chef, who apologized profusely and sent out new versions of the dishes, this time covered with so much genuine saffron that they were barely edible.

Restaurateurs only get away with this because they know so many diners would not recognize the authentic taste of saffron—pungent and ethereal at the same time—or would be too embarrassed to complain, even if they spot the mistake. Buying fancy food gives everyone an incentive to pretend that things are just what they should be; no one wants to look as if they are out of their depth. Of course, not everyone likes genuine saffron; I prefer my paella white, untouched by the stuff, and would rather eat raisin toast than saffron bread any day. But if saffron is what you are paying for, then saffron is what you should get.

There are many gourmet foods whose rarity seems to offer an irresistible invitation to swindlers. Périgord truffles are known as "black diamonds" on account of their value and scarcity. To connoisseurs, they have a glorious aroma, fungal and rich. Only around 120 tonnes of genuine Périgord truffles are now produced annually, fetching prices of up to 3,500 euros per kilo. Yet as many as 300 tonnes of "Périgord truffles" will be sold each year. Many of them are actually inferior black truffles from China, whose true market price is a mere 25 dollars per kilo. These seem to be especially common in the tourist restaurants of the region; the swindle wouldn't work on the locals.

Compared to the genuine article, Chinese truffles have a rubbery, unrewarding taste, with a bitter edge. To the old truffle hunters of Périgord, these counterfeit truffles constitute a fraud not just on the consumers who buy them, but on the producers of genuine truffles. Their real fear comes from the thought that the ignorant consumers won't complain; instead, the worst that could happen is that one taste of a Chinese truffle might be enough to put diners off real French truffles for life.[21]

Such fears have propelled the recent proliferation in Europe of "protected origin" foods. The French system of Appellation d'Origine Contrôlée, or AOC, makes it illegal to sell one of the protected products unless it complies with strict rules of geography and quality. AOC began in 1919 when France passed a law for the Protection of Place of Origin. At first, the AOC status was mainly granted for wines (as we saw in chapter 2), in order that the essential *terroir* of a Rhône wine, say, should not be diluted by unwanted imitators. Increasingly, though, it has been applied to cheeses, too. Roquefort was granted AOC status as long ago as 1925, but most AOC cheeses have been more recent: Crottin de Chavignol (1976); Brie de Meaux (1980); Camembert (1984); Époisses (1991); Banon (2003). More recently still, the AOC label has been granted to many of the other foods of France, products that have a special relationship between place and taste: the honey of Corsica; the chicken of Bresse; Bayonne ham; sweet onions from Cévesses; chestnuts from the Ardèche; real black olives from Nyons; nuts from Grenoble. This protected status sets rigorous standards for the method of production and makes it harder for counterfeiters to cheapen the food.

Despite its modern genesis, the AOC system is at heart the same kind of scheme as the medieval guild networks. Like the food guilds, it starts from the position of wanting to defend the quality of one particular food against those who might debase it. Knowledge is mainly held by the producers rather than the consumers, and it takes the form

Ossau-Iraty cheese bearing the AOC mark—one of a growing number of
traditional European foods to be given protected status.

of local expertise. Information overload is kept to a minimum. All the
consumer needs to know is that a food with a genuine AOC label will
never fall below a certain quality, because the producers know what
they are doing. Like the guilds, AOC can seem rigid and arbitrary in
the way it dictates what can and cannot pass the test. Critics would
say that food producers manage the system, complacently, to their
own advantage and are resistant to newcomers. Where is the room for
culinary innovation? On the other hand, AOC really works.

The AOC system makes it harder for sellers to use place names for
food as if they were generic, as if place of origin had no bearing on
taste. In the early 1990s, there were often "Puy lentils" for sale that
came from Canada. These were perfectly nice, green lentils, the size
of small pills. They were usually labelled perfectly clearly as from
Canada. No deception, perhaps, was intended. But true *lentilles de
Puy* they were not. Lentils have been grown in the Puy region in
the Auvergne for two thousand years. They have a slate-blue colour
and a nutty, mineral taste that is all their own. They seem to retain
their shape and texture better than most other lentils. In 1996, they
became the first legume to be awarded an AOC mark, and since then

more consumers have become aware of their special qualities. This is a good thing. Canadian green lentils are still for sale, but they can no longer be called "Puy" (though some of them are now labelled "*Du Puy*," in an all-too-familiar sleight of hand).

The success of the AOC system in protecting food quality has inspired imitators. Italy has a system of *Denominazione di origine controllata*; Spain has *Denominación de Origen*. More broadly, the European Union has a system of Protected Designation of Origin (PDO) that covers hundreds of special foods. To read the lists of protected fruits and vegetables is to recover a sense of the poetry of particular foods from particular places, far away from the standardized cotton-wool apples lamented by George Orwell: the red radicchio of Treviso; melon from Quercy; Calasparra rice; Calanda peaches; Jersey Royal potatoes; Jumilla pears; asparagus from the Landes; Agen prunes; pink garlic from Lautrec; Aegean pistachios; gherkins from Spreewald; Sicilian blood oranges; Marille (a kind of apricot) from Wachau; Portuguese chestnuts from Padrela; San Marzano tomatoes; Sorrento lemons; Roman artichokes from the Lazio. Such foods deserve protecting; they embody the qualities that make food a pleasure and an art, rather than mere sustenance. Swindling flourishes when people are ignorant of what a given food should really taste like. The PDO system spreads knowledge of the variety of food. It teaches that Prosciutto di Parma is not just any cured ham, but a particular ham with a melting rosy texture made from pigs fed the whey left over from Parmesan cheese; it is not the same as Serrano ham or San Daniele ham, though these too are protected.

In some cases, the protection of particular foods against fraudsters has been taken to extremes. In Corsica, there has long been a problem with swindlers copying the famous charcuterie of the island. Corsican pigs are the main reason that Corsican hams and dried sausages taste so wonderful. The pigs, who live for a good two years, feed on hardy herbs, such as lavender, rosemary and thyme, and on chestnuts and acorns. This diet gives their flesh an herbal depth, a savoury nuttiness, making the local charcuterie sublime. The pigs are carefully restrained from feeding on olives or beech acorns, which would make their flesh greasier. Inevitably, the tourists who flock to rugged Corsica are keen to sample the authentic charcuterie of the region. What many of them end up buying is actually mass-produced pork flown in from China. In 2003, Paul Deminati, a promoter for Corsican charcuterie, commented that "Our charcuterie is known all around the world. When a product is so well known, it always attracts people who make counterfeits. The taste is not the same, but the price is."[22] The response of the Corsican pork industry has been to start a campaign to identify the true race of Corsican pig, tracing the bloodline of each pig back for three generations. The personal history and pedigree of each pig will be stored on a computer. The idea is that only pure-blood, certified Corsican pigs with the right genes will be allowed into Corsican sausage, making it easier to prosecute swindlers.

Despite the use of computer technology, this tracing of bloodlines is no more than another medieval bulwark against counterfeit modern food. The Corsican pig farmers are trying to assert the case that breeding and *terroir* count for more than bland nutritional information. What they are doing is conservative and backward-looking; and rightly so, since the Corsican pigs lead happier lives and taste better than the Chinese interlopers. Elsewhere, however, DNA testers are trying to bring the art of food detection into the twenty-first century.

The DNA of Basmati Rice

Dr. Mark Woolfe is an Arthur Hill Hassall for our times. He leads the Authenticity Unit at the British Food Standards Agency (FSA) in Holborn, which is the main body scrutinizing the activities of food fraudsters in Britain. One of thirty-odd programmes at the agency (which was set up in 2000, in the wake of the BSE crisis), the Authenticity Unit deals not with toxins, additives, or contamination but with cases of deception, misdescription, and mislabelling—cheating, not poisoning, in other words. Authenticity is the term all scientists now use when talking about food that is not adulterated. Authentic food in this sense does not mean ravioli cooked exactly as a Sicilian peasant would make it. It means food that does not cheat anyone.

Woolfe is a tall, imposing man but with the gentle, quiet manner of a true scientist. He shares Hassall's hatred of cheats, and his reverence for scientific method above all else. He could happily talk all day about the minute points of DNA testing and often does. His office has a whiteboard on the wall, where he scribbles formulas. There is an African violet on his desk. Woolfe, who has been at the FSA since it was set up in 2000, is proud of his coffee, which he makes for himself in the office using a small cafetière and freshly ground beans. Like Arthur Hassall, he takes a keen interest in the purity of coffee. Back in 1994, when the Authenticity Unit was part of the Ministry of Food and Farming, Woolfe worked on a survey of the quality of instant coffees for sale in Britain. He found that 15 percent of the samples, mainly the cheapest coffees, turned out to have suspiciously high levels of sugars (xylose, glucose, and fructose), indicating that they had been adulterated, perhaps with coffee plant husks or with maltodextrins.[23] Most of the big-name brands of instant coffee were found to be OK. But Woolfe is taking no risks, sticking to the real stuff. He prefers the taste, anyway.

Woolfe's job involves both surveillance and research. His working party undertakes surveys on different foods, using the latest scientific research to uncover whether given foods on the market are "authentic"—in other words, if they are what they say they are, and of a standard that consumers would expect. These surveys are time consuming, and only a handful are conducted each year. The process can be painfully slow. Woolfe says that his team has to prioritize carefully which foods to survey. The decision reflects consumer concern but also the value of a given market; they are more likely to look at meat products, where potential fraud is worth "billions of pounds," than honey, a market that is relatively small. "We're a hidden force," he tells me. "It's all to do with consumer confidence. People should have confidence that when people buy something it is what it says it is." It could be Arthur Hill Hassall speaking.[24]

Like Hassall, Woolfe sees the importance of revealing when fraud has not taken place as well as when it has. Hassall demolished the myth of the grocer who put sand in his sugar. Woolfe has busted a few myths of his own. In several cases, the Authenticity Unit has shown that consumer fears about adulteration were unfounded. There have long been rumours in Britain about donkey meat and horsemeat finding their way into salamis and other cured meats from the Continent. These fears were partly xenophobic, based on the notion that foreigners eat suspicious things. As Woolfe himself comments, "anecdotally, people thought things coming in from central Europe had all sorts of nasties in them." In 2003, the Authenticity Unit did a survey of 158 salamis and "salami-like products," looking for undeclared horsemeat and donkey meat. They discovered that this kind of fraud is practically nonexistent. There was only one trace of horsemeat in a single sample of chorizo.[25] It is not in fact as if horsemeat or donkey meat salamis are necessarily bad things: "They are very low fat meats," as Woolfe rightly observes. The point, though, was that consumers believed, wrongly as it turned out, that they were being tricked into eating these meats in place of pork or beef. Such feelings of unease around food are bad for everyone—producers and consumers alike—because they erode trust.

In other cases, Woolfe's Authenticity team has shown consumer fears to be right, and in so doing, addressed those fears. It is a commonplace that modern meat and fish is pumped full of water, deceiving us into paying more for it, weight for weight, than we should. Sometimes "added water" is declared on the label, but even so, the suspicion remains that the wool is somehow being pulled over our eyes. You suspect that you are being cheated, but there is little you can do about it. Sure enough, an FSA survey of frozen chicken breasts from 2001 uncovered a major problem of excessive added water—water that remained in the chicken even after it had been thawed and cooked, because it was held in by the use of hydrolysed proteins of pork and beef. Most of this "injected" chicken was produced in Europe and was destined for the restaurant trade. This was a scandal, and a twist on the ages-old swindle of padding food with water, because of the presence of pork protein. Some of this porky water had been injected into chicken labelled "Halal"—a particularly unkind deception, which turned cheating into poisoning for Muslims who ate the affected chicken. Equally startling was Woolfe's survey of 2002 of added water in raw scallops and scampi. A scandalous 86 percent of the "ice-glazed" peeled scampi consisted of more than 10 percent added water. Of the samples of scallops, 48 percent contained more than 10 percent added water; in the most extreme case, there was 54 percent added water, more water than scallop, a truly audacious piece of fraud; if there had been any less scallop, it would have been little more than an ice cube, all sold at a premium price as finest "king scallop."

There is no doubt that these surveys have done good. The most blatant cases of cheating have been prosecuted by the relevant local authorities. Meanwhile, the food industry in general has started to clean up its act in response to these reports, usually cooperating with the government's requests for more stringent codes of practice. It is hardly something to boast of, to be named and shamed as the manufacturer of the wateriest frozen scallop in Britain (that particular honour belonged in 2002 to Colncrest, a wholesaler in London, though Ice Pak International in Leeds ran a close second). From Woolfe's point of view, the success of this kind of surveillance is entirely dependent on having a "robust scientific method" to start with. As with Hassall and his microscope in the 1850s, Woolfe needs to be sure that his method is sound enough to offer the most accurate information to consumers, as well as to forestall legal action from manufacturers.

The battle is still the same. The two sides, the goodies and the baddies, still progress at roughly the same rate as regards what science can either conceal or reveal in food. The swindlers still have roughly the same kind of expertise to draw on—legal and scientific—as the antiswindlers. The burden of proof on the antiswindlers thus remains extremely high.

One thing has changed, though. The science of "authenticity," at the level of both detection and deception, has been transformed since the 1850s. There will always be room for organoleptic tests—those basic techniques of sniffing, examining, and tasting—but if they are to stand up as "evidence," they must be backed up with sophisticated scientific experiment. The microscope, which did such sterling service for Hassall, has been largely superseded. It is still used occasionally when analysing such things as fruit jams and purees—under the microscope, it is easy to see if pears or rhubarb have been used to bulk out expensive berries, because the hard, stonelike cells of

289

pears and the long, fibrous cells of rhubarb differ from the soft fruits. Mainly, however, microscope work has been replaced by gas chromatography, isotope analysis, and mass spectroscopy, among other methods.

Gas chromatography is a series of techniques for separating mixtures, where a sample is vaporized and passed through an inert gas in a column; a "detector" is then used to reveal the makeup of the substance, which comes out on a moving graph, like a heartbeat. Chromatography is an excellent tool for detecting the rising fraud in olive oil.[26] The popularity of the "Mediterranean diet" has led to increased demand for Extra Virgin Olive Oil (EVOO, as some chefs now insist on calling it), which has given rise to the practice of diluting it with bog-standard refined olive oil. Armando Manni, who produces some of the best olive oil in Italy, has complained that much of what passes for "virgin olive oil" is no better than "lamp oil," repackaged with fancy labels. Chromatography can detect this fraud, by isolating the sterol content of a given sample of oil. Refined oils have much higher levels of erythrodiol than do virgin oils.

For foods whose identity is closely linked to a particular environment, the best tool for establishing authenticity may be isotopes. "Locked in every plant and animal is a chemical memory of the weather and environment it grew up in."[27] Like the rings on a tree telling us about the plant's age, isotopes can give us information about where and how a food was produced. Isotopes are different versions of the same element, with different atomic weights. By looking at the isotope ratios of the biological elements (carbon, hydrogen, oxygen, and nitrogen) in a given food using mass spectrometry, it is often possible to say where it came from or how it was produced. For example, you would expect to see a lot of lightweight hydrogen isotopes in English lamb and a lot of heavier hydrogen isotopes in Spanish lamb, because of the different chemistry of the ground or rain water drunk by the animals in the two countries. Isotopes can reveal what food a chicken has eaten (is it *really* corn-fed, as the packet claims?), whether salmon is wild or farmed, whether Parma ham truly comes from Parma, and whether honey from a particular area of the world has been adulterated with sugar or corn syrup.[28]

Isotopic measurements have also been used to detect the common scam of bulking out fruit juices with water, sugar, and pulpwash, a liquid extracted by repeatedly washing the exhausted pulp left over from making proper juice. In the early 1990s, it was calculated that fraudulent juices made up as much as 10 percent of the U.S. market. In Britain, a study done in 1991 found that sixteen out of twenty-one leading brands of orange juice contained added substances, mainly beet sugar, whose isotopic composition is similar to that found in oranges. Spectroscopy—which measures the molecular mass of a given sample—can show peaks in the sugar level of orange juice. By itself, however, this tool may not be enough. After all, oranges in nature vary in their sugar content. Food analysts may need to combine a number of techniques to come up with the "exquisitely sensitive assessment" they need.

One of the most "exquisitely sensitive" of all current tools is the use of DNA, which Woolfe's programme has pioneered. He compares this kind of "food forensics" to "the DNA profiling of criminals," the difference being that profiling food is "much more difficult."[29] With food, it is not the criminal but the crime that is being profiled, and the method has to be reworked for every different food. DNA analysis doesn't work for everything—it is not much use for uncovering the kind of chemical mixtures exposed by Accum—but it is outstandingly good at detecting any fraud to do with species. As Woolfe has written, "DNA has the discriminating power because ultimately the definition of a variety or species is determined by the sequence of its genome." This is useful because this kind of fraud is on the rise. Discerning eaters increasingly choose which foods to buy based on specific strains and breeds—Cox's apples, Gressingham duck—which means that sly swindlers will increasingly try to pass one variety off as another. Now, the genetic "fingerprint" of different apple varieties can be recorded, making fraud harder. In 2003, the FSA did a survey of potato varieties. The British shopper will pay more for certain types of potato, such as King Edward, which are the classic floury English potato with a creamy interior, or Charlotte, which are yellow and waxy. A total of 294 samples were obtained, of which 33 percent were mislabelled and 17 percent were not the variety of potato they claimed to be. Many of

291

the "King Edward" potatoes were actually Ambo potatoes, a variety that sells for half the price. It was possible to obtain this information only by doing a molecular genotyping of the relevant potatoes. The technique is so powerful, it can distinguish between fifty different commercial varieties of potatoes using only five DNA markers.

Similarly, the agency's programme has come up with a DNA method for detecting the deliberate adulteration of durum wheat pasta with bog-standard common wheat. This could become a major problem throughout Europe, since the price differential between the two types of wheat is large. Proper dried pasta should be made using only durum wheat, whose hardness gives pasta its lovely al dente bite. Ordinary wheat flour makes flabby, sticky pasta, but it is much cheaper than durum wheat, so falsified pasta is common, especially in the lowest price range. Previously, there had been no accurate method, other than tasting, for checking for the presence of common flour in pasta. Scientists funded by the agency have worked out a way of amplifying a small sequence of DNA on the D-genome of common wheat; this genome is lacking in durum wheat. With this "assay," it should now be possible to uncover even small additions of wheat flour to pasta. Elsewhere in Europe, researchers have used DNA testing to detect the presence of cow's milk in "buffalo mozzarella" and in sheep's milk cheese.

As for Woolfe, his greatest DNA triumph has been Basmati rice. It is generally acknowledged that Basmati, grown for hundreds of years in the foothills of the Himalayas, is the finest long-grain rice you can buy. If you want to make the best pilau or biriyani, then Basmati is what you need. It is the most delicate and fragrant rice (the name Basmati comes from the Hindi word for fragrant); even when plainly boiled, it is a feast. Apart from its perfume, another special property of Basmati is its long, thin grain; when you cook it, it almost doubles in length, due to the special character of its starch. Inevitably, it is more expensive than ordinary long-grain rice, both on account of its superior qualities and because it is difficult to grow; Basmati plants are temperamental and low-yielding. The cheapest Basmati retails in British supermarkets for around £1 per kilo, compared with 50p for basic rice; the finest, brand-name Basmati costs more than £2 per kilo.

Genuine Tilda Basmati rice. When the British Food Standards Agency surveyed the authenticity of Basmati rice on sale in Britain in 2003, Tilda rice was found to be pure; many others were not.

In the 1990s, Woolfe and his colleagues (then working at the Ministry of Agriculture, Fisheries and Food, before the Food Standards Agency was established) suspected that there might be a problem with Basmati being padded with inferior rice varieties. The trouble was, there was no technology equipped to find out the truth. Because of Basmati's irresistible perfume, Woolfe looked into the possibility of characterizing it using an electronic nose. Could adulteration be detected by getting such a "nose" to analyse the smell of the cooking water left behind when Basmati was boiled? This proved a dead-end for detecting mixtures. It turns out that different batches of Basmati, even when entirely genuine, have different strengths of aroma. Some just have a naturally weak scent.

Then, in 1999–2000, scientists working at the University of Nottingham came up with a technique that Woolfe could use, employing similar DNA markers to those used when doing genetic "fingerprinting" of humans. The technology, known as PCR—short for polymerase chain reaction—uses enzymes to amplify a particular string of DNA sequencing.[30] This gives a genetic snapshot of a given plant or animal. By looking at the patterning of a short DNA sequence, it is possible to distinguish one species from another, provided the right sequence is found. In the case of Basmati, the scientists at Nottingham screened a large number of DNA sequences before eventually narrowing it down to twelve markers, which could distinguish even closely related rice cultivators from each other with accuracy. The "assay" was also made quantitative, so that it could measure not just whether Basmati was adulterated, but how much non-Basmati rice was present in a given sample. At last, Woolfe had the "robust method" he needed to test how much Basmati swindling was going on.

There was still the problem of defining Basmati, however. Unlike the special PDO foods of Europe, the name "Basmati" has no protected status. In UK law, Basmati was simply a "customary name"—a name that consumers can supposedly understand without need of further explanation. This may be true in the *mandis* (or markets) of India and Pakistan, where wholesale buyers are so knowledgeable

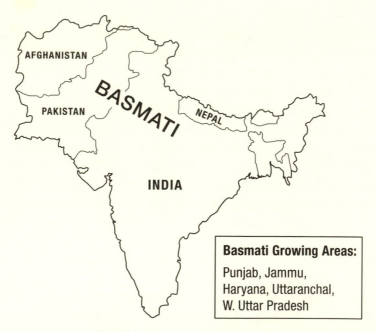

Basmati Growing Areas:

Punjab, Jammu, Haryana, Uttaranchal, W. Uttar Pradesh

Map of the Basmati rice growing areas.

about the different grades of rice that they can identify the various strains of Basmati simply by chewing on a raw, unhusked grain, gauging its exact flavour. It is not true in the supermarkets of the West, where most ordinary consumers have little sense of what the word on the label means, beyond the fact that Basmati rice tastes nice.

Woolfe's starting point was the generally accepted place of origin of the rice, which grows in the foothills of the Himalayas—in the Punjab, Haryana, and Uttar Pradesh regions of India, as well as the eastern Punjab of Pakistan. He soon found, though, that to attempt to pin down exactly which strains of rice count as Basmati was to stumble into the minefield of Indian-Pakistani politics. The Indian authorities recognize only six traditional, "true-line" strains (Basmati 370, Dehra Dun, Basmati 217, Basmati 386, Taraori, Ranbir Basmati). They differentiate these from the evolved or hybrid Basmatis—Pusa Basmati, Punjab Basmati, Haryana Basmati, Kasturi, Mahi Suganda. Since the Indian export trade to Europe is mainly in traditional varieties, they favour these over the newer hybrids. The Pakistani authorities, by contrast, recognize five varieties of Basmati, their main trade being in three modern hybrid varieties—Super Basmati, Basmati 385, and Basmati 198. So while the Indian authorities see hybrids as interlopers, in Pakistan they are "authentic." Given the political friction between India and Pakistan, these seemingly trivial botanical preferences can become loaded with menace.

Woolfe, a mild man, was alarmed at having stumbled into this minefield of Indo-Pakistani bubbling tensions. In the end, for reasons of diplomacy as well as to avoid distorting either country's trade, the FSA decided that it must recognize all varieties that had been approved by both India and Pakistan—the modern Pakistani varieties as well as the older Indian ones, with the proviso that they must have been bred from at least one "parent" of a trueline Basmati variety. All the recognized varieties needed to have the special long, thin Basmati grain, and that inimitable fragrance.

Now equipped with both a method and a definition, in 2003 the Authenticity Unit organized a survey through local authorities who gathered 363 samples of Basmati rice from shops all over Britain.

The findings were depressing if not surprising: only 54 percent of samples were pure Basmati rice; the others all had cheaper rice mixed in. Thirty-one samples contained non-Basmati rice in excess of 60 percent—which means that they hardly contained any genuine Basmati at all. Given the price differential between Basmati and ordinary long-grain, someone was getting very rich from the adulteration. It was calculated that in 2002 alone, Basmati consumers were swindled out of more than £5 million. As Woolfe comments, "It wasn't just UK consumers being cheated, but also EU tax payers being cheated." For reasons not altogether clear, at the time of the survey, Basmati rice was subsidized as it entered the EU, to the tune of up to 250 euros per tonne. Thus, importing non-Basmati rice as if it were Basmati was a hugely profitable business.

When Woolfe's survey of Basmati rice was published in 2004, listing the names and addresses of the suppliers, the industry was shaken. Two Essex-based companies were fined more than £8,000 each for selling "Basmati" that had been adulterated with between 55 and 75 percent non-Basmati rice; it was not so much the sum of money as the exposure that stung. More widely, the new surveillance has enabled a much more rigorous code of practice for the industry. The previous code of practice allowed for Basmati legally to contain up to 20 percent non-Basmati rice; this has been revised to 7 percent, still higher than Woolfe would like, but, considering the basic agricultural and handling practices in the countries of origin, it is a step in the right direction.

Now that they are aware of the existence of DNA testing methods, sellers are much warier about palming off non-Basmati rice as Basmati. They have also seen the benefits of selling the real thing: the status of Tilda Basmati rice was only enhanced by the survey's confirmation that it was 100 percent Basmati. A spokesperson for East End Foods, another manufacturer to come out well, commented gleefully, "I feel this is one of the best projects the FSA has undertaken." Meanwhile, the Basmati growers of India, who are mostly still smallholders who farm in the old-fashioned ways, have been emboldened by the new knowledge of Basmati's DNA to petition the World Trade Organization for protected status for the name. This would help to arm the Indian growers against the threat of unwanted competition from new U.S. hybrids sold under the trade name of "Texmati," a cross-breed of Basmati with American long-grain rice, which claims to be "the most widely recognized brand of aromatic rice in the U.S."

It is easy to see why Woolfe says that "Basmati is one of our really big success stories," a touch of delight in his voice. At the same time, there is little chance of him resting on his laurels. Already, in the few years since the DNA-testing method was developed, unscrupulous rice traders have changed their tactics to evade detection. In 2003–4, the main adulterants used in Basmati mixes were the non-Basmati varieties Sherbati and Pakistan 386. The agency's "assay" was purposely designed so that these varieties could be easily distinguished from true Basmati. As a result, in a sly piece of biopiracy, some merchants have switched to Yamini and Pusa 1121, new hybrids that are genetically very similar to some of the approved strains and are much harder to pick up using the current DNA markers. New markers will have to be developed to take this into account; not impossible, but costly. By the time the new method is ready, the science of swindling may have moved on too, with yet more varieties of rice bred to fool the detectors. "It's a moving beast," says Woolfe. "The minute you come up with a method, people find ways around it."

The best that Woolfe can hope for is that the method will do as much good as possible before it becomes obsolete. For this, the method needs to be disseminated far and wide and backed up with enforcement. Hassall was one of the first public analysts—the body of scientists who analyse suspicious food samples at a local level. One of Woolfe's achievements has been to ensure that new DNA methods reach the public analysts so that they can be used in routine policing of the food supply. The original DNA assay for Basmati needed expensive equipment, which many public labs simply couldn't afford. Woolfe's Authenticity team therefore worked to create a series of cheap, portable "labs-on-a-chip." Using fine capillaries on a glass plate chip, rice samples are passed through three different colours of enzyme gel, which show different repeat sequences of DNA. By selecting the appropriate markers, these mini labs can easily and accurately differentiate between the approved Basmati varieties and the fakes. Each test only costs a few pounds and can be performed by a non-specialist. The same technology is being adapted by the Authenticity programme to create mini labs for revealing different fish species and meat species, as well as the adulteration of fruit juices.

The quiet, behind-the-scenes food forensics of Mark Woolfe shows that in some ways things have not changed since the days of Accum. The battle against adulteration is still the fight of the science of detection against the science of deception. The battlefield has moved, though, from chemistry to biology; and from deceptions that were easy to explain, once exposed, to those so complex and technical that consumers may never fully realize exactly what they have been duped with or rescued from. Insofar as they *are* aware of food regulation, they may resent it, as an unnecessary intrusion. Food dangers that cannot fully be explained to the layperson raise their own perils. We risk creating a populace that moves from apathy to hysteria to apathy again, without ever developing what Mark Woolfe himself seems to enjoy: a cool appreciation of the blessings of eating authentic food.

Invisible Dangers and Scaremongering

In the nineteenth century, as we have seen, food laws were responding to threats that were terrifying and direct. There was nothing theoretical about the arsenic-laced lozenges that killed twenty people in Bradford in 1858.[31] Modern dangers from food tend to be different. In the words of a lawyer, "Harm has become ephemeral and distant, not hard and immediate."[32] The threat of carcinogenicity, hovering over so much food, is almost impossible to connect with the daily act of eating. If you ate an arsenic lozenge, you keeled over more or less instantly. Now, you may eat thousands of supposed "carcinogens," from burnt toast to pesticides and additives, over many years before you develop cancer; or you may be lucky and never get ill at all. There is another difference too. Then, food fear cropped up in specific epidemics of deceit. Now, it is so omnipresent it is comical.

In this atmosphere of universal mistrust of food, it becomes easier to be gulled by scaremongers who manipulate the fear of food to sell their wares. There is no shortage of food shamans—some of them call themselves "nutritionists"—prepared to tell you that you will irrevocably damage your body if you take another bite of cooked food/wheat/dairy/anything really, other than their own special approved food supplement, which you can conveniently send off for, only $69.99 for a month's supply plus postage and handling.

Other food scares become exaggerated not because anyone stands to gain by them, but because they take on a life of their own, spiralling out of control. The Sudan 1 affair of 2005 was just such a needless panic. The biggest food recall in British history was triggered after some adulterated chilli powder was found in Crosse & Blackwell Worcester sauce, which had been supplied to all the main food manufacturers, going into more than four hundred supermarket products, from chicken lasagne and pepperoni pizza to country vegetable soup and shepherd's pie. The discovery created widespread alarm. It was found that the chilli powder, imported from India, had been touched up with a red "azo dye" called Sudan 1, which has been declared a possible carcinogen (though this has not been proven), and which is commonly used in floor polish and boot polish. People were disgusted to think they had been eating cancer-causing boot polish! The Food Standards Agency tried to allay public concerns—"no need to panic" was the message from the FSA chairman. Yet the agency itself exacerbated the panic by demanding the recall of all four hundred affected products. Shoppers were alarmed. Usually such a gesture would be reserved for a serious hazard, such as food contaminated with shards of glass. What was forgotten in the midst of the panic was that the amount of Sudan 1 that had made its way into the processed foods was so negligible as to pose no risk at all to human health. Moreover, as recently as 2003, it was perfectly legal to import foods containing Sudan 1 into Britain. Thus only two years earlier, Sudan 1 had been an "additive." not a toxin. The whole scare was a phantom.

Just as false and exaggerated dangers swarm everywhere, real dangers may become invisible. Trans fats (the popular name for partially hydrogenated fats or trans-fatty acids) are now generally seen to be something you wouldn't want to eat too much of. Hydrogenated fats are produced through a chemical process that hardens vegetable oil

299

by passing hydrogen through it. There is incontrovertible evidence that consuming too much of them can contribute to high cholesterol and heart disease. In December 2006, New York City banned all but tiny amounts of trans fats from the city's restaurants.[33] Big food corporations from Starbucks to KFC have announced that they are cutting back on trans fats in the foods they sell.

The question remains, however, what trans fats were doing in food in the first place. Their risks have been known about for thirty years.[34] Yet until very recently, trans fats found their way into as much as 40 percent of "ordinary" processed foods, lending crispness and long shelf-life to cakes and biscuits, breakfast cereals and breads. It was only after an enterprising lawyer sued the manufacturers of Oreo cookies over their high trans fat content in 2003 that the food industry was finally shamed into looking for healthier fats. Before that, despite knowing the risks to the poor saps who bought their products, food manufacturers adored trans fats because they were cheap and reliable. Consumers were kept in total ignorance of their presence, since there was no requirement to list them on the label, alongside saturated and unsaturated fats (this changed in the United States as of 2006).

Worst of all, many products high in trans fats might say that they were "low in cholesterol" or "low in saturated fat," statements that were true but entirely misleading, since trans fats are probably just as unhealthy as saturated ones. A couple of years ago, while researching an article on trans fats, I phoned the customer care line of a major British biscuit manufacturer. I told the "customer care adviser" that I was concerned about the fat in their biscuits: was it hydrogenated? A kind-sounding lady advised me not to worry; the fat was only "*partially* hydrogenated," the implication being that this meant that it was only *partially* bad. (Perhaps, when my time came, it would give me only a *partial* heart attack.) The line that this nice lady had been asked to spin by her bosses was abject nonsense. In fact, all trans fats are "*partially* hydrogenated"; when the fatty acids become fully hydrogenated, they are free of trans. A lot of this kind of active misinformation must have gone on before the public finally woke up to the dangers of trans fats; before we moved from apathy to hysteria.

Trans fats were not the first invisible danger to be exposed as "not on the label"; nor will they be the last. Food campaigners have increasingly turned their attention not just to what is in food, but how it is produced. Processing aids are one of the hidden scandals of modern food, because unlike additives, they do not have to be declared on the label. Take enzymes. While Mark Woolfe uses his enzyme gels to detect adulterations, other enzymes are being used to disguise the true properties of foods. Enzymes are used to tidy up many of the inconvenient defects of nature: there are enzymes to tenderize meat and ripen cheese, enzymes to prevent discolouration in shrimp or white wine; enzymes to keep pickles crisp and to peel mandarin oranges before they are canned; and, most of all, enzymes to help make the perfect modern industrial loaf of bread: soft, white, and slow to go mouldy.[35] To the baking industry, enzymes are "tiny invisible helpers," proteins that speed up biological reactions, enabling industrial bakers to make stretchier dough, or to make high-rise bread from poor-quality flour, or to knead in more water, thus boosting profits.[36] Most consumers have no idea they are even there. The reason that processing aids do not have to be listed on the label is that they are supposedly used up during production, leaving no trace of their existence behind. Yet this is contradicted by the fact that enzyme manufacturers themselves boast about the "thermo-stability" of their products, their ability to withstand the heat of the oven without altering; and by studies that have shown that significant traces of enzymes can remain after baking, including amylase, which is known to cause serious allergic reactions in some people.[37]

Enzymes in bread may seem like a trivial problem compared to arsenic lozenges. But the unheralded presence of these substances in bread is part of a bigger problem of food ethics. As the artisanal baker Andrew Whitley says: "If you don't know what's gone into your bread—in the fullest sense of those words—how can you exercise any meaningful choice over whether to eat it or not?"[38] The European law on processing aids is currently under review; the Food Standards Agency is also looking at tighter labelling laws for enzymes. In the future, enzymes may have to be declared on the label, but they will probably be listed under the blanket term "enzymes," with no indication of which particular enzymes have been used. For Whitley, this is not enough. "People have a right to know not just what is in their food but how it has been made."[39] This right has become all the more pressing in an age when the techniques by which a food is reared or grown have come under as much scrutiny as its ingredients. While much of the story of swindling in the twenty-first century turns out to be familiar, here is something genuinely new: the threat of biological changes to the basic structure of food itself.

Fatty Chickens and Agricultural Adulteration

At every other period in history, the best way to avoid adulteration has been to eat food that was what it said it was: basic agricultural produce. Arthur Hassall wanted mustard that really was mustard. Harvey Wiley sought out preservative-free meat. "EAT WHOLE, FRESH FOOD" was Caroline Walker's summation of a lifetime of giving nutritional advice. The swindlers might be able to tamper with mixtures and powders, but some foods had an integrity that couldn't be tampered with. A carrot was always a carrot; a chicken was always a chicken. If you stuck to fresh vegetables and fruits and poultry and wholewheat bread, you would be all right. Now, though, these old certainties seem to be crumbling, as the essential values of basic foods have changed.

In 2004, chicken—a supposedly low-fat, "healthy" meat by comparison with beef—was found to contain nearly three times as much fat as it had thirty-five years earlier, thanks to changes in farming methods.[40] Professor Michael Crawford, an expert on fats who works at the Institute of Brain Chemistry and Human Nutrition at London Metropolitan University, commented that consumers who bought modern chicken in the belief it was good for them were "being sold a pup down the line."[41] For the first time since records had begun in the 1870s, the fat content in a basic roast chicken had overtaken its protein content. "Six times as many calories are coming from fat as from protein," noted Professor Crawford. This was the case even with chicken that had been plainly roasted with lemon and salt, without so much as a drizzling of oil. In 1970, chicken contained 8.6 grams of fat per 100 grams; now, the average supermarket chicken contained a whopping 22.8 grams. A single roasted chicken leg had more fat than a Big Mac. The British now eat close to 30 kg of chicken per person per year, more than twice as much as in the 1970s.

Even if you avoid the skin—the fattiest part of the bird—and eat only the white breast meat, you still cannot get away from the essential problem. Crawford insists that "This whole focus on rapid growth, achieved through a high-energy, high cereal-based diet has changed the lipid composition of the chicken meat itself."[42] The chicken has been adulterated from within. Advocates of better animal husbandry would agree. The life of the average broiler chicken, debeaked, crammed in darkened, filth-encrusted sheds, and fattened up to a physique so top-heavy it could not fly even if it was given the chance, has been debased to a point where it is doubtful whether it should be called "chicken" at all. All this has been done to satisfy consumer demand for healthy white meat; there is a certain bleak justice in discovering that the broiler system is bad for us as well as for the poor chickens.

Human physiology is adapted to eating wild meats, which are dense and lean. As Crawford says, "Chickens used to roam free and eat herbs and seeds. . . . You just wouldn't find anything like these [supermarket] chickens in the wild."[43] When modern industrially produced meat is put under the microscope—beef as well as chicken—it is possible to see signs of "pathological fat infiltration," a fat so pervasive it inveigles its way between the animal's muscles. The muscles themselves have largely atrophied from lack of exercise. Crawford argues that of the chickens currently on the market, "a free-range bird is best," one that has been permitted the freedom to exercise its limbs and some time to grow at a natural pace; but even then, the fat content would depend on the diet the bird had been fed. Essentially, Crawford's findings on chicken meant that "we now need a new definition of what we mean by a healthy food."[44]

Some would argue that the same is true of modern fruits and vegetables, traditionally the last bastion against adulteration. Tim Lang, professor of food policy at City University in London, has said: "We think of something like an orange as a constant, but it isn't."[45] In 2002, a Canadian newspaper reported that fruit and vegetables sold in supermarkets contained "far fewer nutrients" than fifty years ago. The report concluded that consumers would have to eat eight oranges to obtain the level of vitamin A that an earlier generation got from a single fruit.[46] Similarly, if records are correct, many British vegetables have lost much of the mineral content they had half a century ago. Broccoli has 80 percent less copper than it once did. Tomatoes have a quarter as much calcium. In each case, blame has been laid on intensive farming methods. David Thomas, a geologist, has claimed that the long supply chains, fertilizers, and hydroponic growing methods of modern agriculture, by which plants are grown not in the soil but in artificial matting, are the cause of this decline.[47]

While some fruits and vegetables have become depleted of nutrients, other have become supercharged with them, a result of genetic modification. Biotechnology has yielded tomatoes bursting with lycopene, vitamin A, and beta-carotene, and Monsanto's "golden rice" enhanced with vitamin A. To proponents of genetic modification, these represent improvements on nature. To critics, GM crops are adulterations, not just of the individual foods that have been altered but of the whole horticultural environment, which may be affected by possible genetic mutation and biosphere damage.[48] At the time of writing, there is still no requirement in the United States that GM foods be labelled as such. Chefs have complained that they may be forced unwittingly to adulterate the food they serve their customers. In 1998, one chef commented that "People come to [my restaurant] because they trust me [and] know I'm going to source out the highest quality ingredients in the market for their dining experience. By not requiring mandatory labelling of all genetically modified foods, the government is taking away my ability to assure customers of the purity of the foods they eat at my restaurants."[49]

Such issues do not arise in Europe, where GM foods must be clearly labelled as such—not that they are often seen, because they are not grown at all in the EU. Even here, however, concerns remain that GM crops could cross-pollinate with non-GM crops and thus "contaminate" them. This is a particular concern with organic food. In 2003, Mauro Albrizio from the European Environmental Bureau commented that "the right to eat GM-free food will be severely compromised if GM crops are grown on a large scale."[50]

Viewed from the opposite perspective, however, GM farming is not the cause of agricultural adulteration but a remedy for it. Bio-tech evangelists argue that by breeding plants that have inbuilt insect resistance—such as genetically modified maize—they will be able vastly to reduce the use of chemicals on crops. Critics dispute this. A detailed study by scientists at Cornell University in 2006 found that GM cotton farmers in China initially reduced pesticide use but ended up using as much as they did with conventional crops.[51] Whatever the facts of the matter, it is interesting that from opposite sides of the great food debate, organic farmers and GM scientists both see heavy pesticide use as a kind of adulteration. They both present themselves as a way of purifying this adulteration.

This is where we arrive at the thorny question of purity once again. It is often forgotten that pesticide use—which some now see as an adulteration in its own right—proliferated partly as a remedy for adulteration. Pesticides once offered the dream of a totally clean food supply.

The "Filth Clause," Pesticides, and Organic Fraud

One of the most troublesome parts of the U.S. Pure Food and Drugs Act concerns "aesthetic adulteration." Clause 402(a) states that a food is regarded as adulterated if it contains any "filthy, putrid or decomposed substance or is otherwise unfit for food." As we saw in chapter 4, the Pure Food and Drugs Act was drafted in the aftermath of Upton Sinclair's The Jungle, and the "filth" the lawmakers had in mind was the kind of unsanitary goings-on Sinclair uncovered among the Chicago meatpackers: diseased beef, doctored sausages, and so on. The lawmakers were not thinking of the clods of soil on an organic carrot; they were not thinking of the decomposed fish in a bottle of Thai fish sauce or the mould on a Stilton. In defining adulteration as "filth," however, this clause encouraged just the sort of squeamishness about food that can send people running away from real food into the arms of industrial manufacturers. As one legal commentator has said, "taken literally," the standard embodied in clause 402(a) "would bar any level of extraneous matter in food and render virtually all food adulterated."[52]

In practice, this did not—could not—happen. The American courts have not prosecuted farmers for selling muddy apples or grapes with the odd tiny insect still running about among the stalks, accepting the fact that eliminating all filth from the food supply is a utopian impossibility. Everyone who works in food production recognizes that no food is entirely clean. The American public, however, has effectively been kept in a state of childlike innocence, its fragile sensibilities preserved from the truth that dirt and food coexist. As Peter Barton Hutt wrote in 1978, "Emotionally, the public is not prepared to accept the fact that agricultural produce is raised in fields and stored in barns."[53] In 1972, in the interests of greater openness about food, the FDA experimented with releasing a formerly secret list of "filth guidelines"—the unavoidable "filth allowances" that existed for each food.[54] Weems Clevenger, an FDA spokesperson, announced that it was impossible to make food 100 percent pure. "There are insect fragments in every loaf of bread," he stated, and an allowance of one rodent pellet per pint of white flour.[55] These revelations were greeted with a public outcry, and the experiment in transparency was not repeated. As Barton Hutt says, "the American public is not yet prepared to face each new day with label statements of the maggots, mould, rodent pellets, rat hairs and insects contained in their fruit juice, cereal, bread, jam and coffee."[56] The same is true in Britain, where an obsession with clean food is pursued under the name of hygiene.

In the United States, filth guidelines still exist, with different action levels for different foods, but they are not widely publicized. A full list is available on the FDA website, but you probably wouldn't find it unless you were looking for it.[57] Reading through the list gives the lie to the notion of perfectly clean food. The average jar of peanut butter may contain up to 30 insect fragments per 100 g and up to one rodent hair; tomato juice may contain 10 fly eggs or 2 maggots per 100 g from the drosophilia fly; ginger is permitted 3 mg of "mammalian excreta" per 100 g; blue-fin fish may have 60 parasitic cysts per 100 fish; fig paste may harbour 13 or more insect heads per 100 g; dried mushrooms are allowed 75 mites per 100 g; ground marjoram may contain an astonishingly high sounding 1,175 insect fragments per 10 g. Some of the guidelines seem amazingly precise: frozen spinach is permitted an average of

50 or more aphids, thrips and/or mites per 100 grams

or

2 or more 3mm or longer larvae and/or larval fragments or spinach worms (caterpillars) whose aggregate length exceeds 12 mm per 24 pounds

or

Leaf miners of any size average 8 or more per 100 grams or leaf miners 3 mm or longer average 4 or more per 100 grams.

It may take a strong stomach to read this, but none of it is really cause for concern. It has been estimated that most of us eat a pound of two of insects every year without our knowledge.[58] They probably do us some good. Philip Nixon, an entomologist at the University of Illinois, has said that insects are "actually pretty healthy," being high in protein and nutrients and low in fat.[59] The *unhealthy* thing is our refusal to face the fact that we are insect eaters. Squeamishness about "aesthetic adulteration" has had a bad unintended consequence: the big increase in pesticide levels, potentially a much more damaging kind of adulteration than the odd fragment of insect. Pesticide usage increased ten times in the United States from 1954 to 1974.[60]

Opinion differs as to how serious a problem is posed by pesticide residues in our food. A survey done by the Food Standards Agency in 2004 found pesticide residues in 31 percent of the food samples tested, but the agency insists that these were mostly at a very low level and do not pose a great risk to health. The agency is keen to downplay the risks involved in consuming pesticide residues since "not eating any fruit and vegetables would be a much bigger risk to someone's health than eating foods containing low levels of pesticide residues." This may be good public health policy, but as individual consumers, we would rather that consuming fruit and vegetables did not put us at risk at all, especially since the efficacy of pesticides in controlling pests has been called into question.

Critics of pesticides point out that entire groups of them are poisonous "often because the mechanism by which they are effective against pests is equally effective against people."[61] They also point to the so-called cocktail effect whereby low levels of many different pesticides taken in conjunction over a long period of time may be more harmful to health than the safety limits for pesticides allow. Having secured your genuine Cox's apple, you may find that it has been sprayed "18 times with many different chemicals," according to the Soil Association, posing a far greater danger than the odd mite.[62] The case against pesticides seems most worrying for infants. Safety levels for pesticides are set based on the bodyweight of an average adult. Relative to their body size, children consume more chemicals from pesticides than adults. In 2005, a study was done testing the urine of preschool children. "The urine of those fed conventional foods contained six times as many pesticide residues as were found in the urine of children who ate organics."[63]

The answer seems obvious: go organic. Though it looks like a modern response to a modern problem, the organic doctrine is really centuries old: the reverie of the self-sufficient yeoman farmer who rejects the toxins of industry in favour of his own home-grown produce. It is the call of Virgil and Rousseau and Thomas Jefferson. It is a vision of removing oneself wholesale from the poisonous city. Organic food only looks modern because the backdrop has changed.

Advocates of organic food argue that as well as lacking the negative effects of pesticide residues, organic produce has positive benefits. A three-year Italian study, reported in 2002 in the *Journal of Agricultural and Food Chemistry*, found higher levels of polyphenols, antioxidant chemicals that may reduce the risk of cancer, in organic peaches and pears than in conventionally grown fruit.[64] A year later, another report in the same journal found 52 percent more vitamin C in organic corn, as against conventional corn, but higher levels still in "sustainable corn," which had been raised by a mixture of organic and conventional methods. In 2006, fourteen British scientists wrote to the FSA calling their attention to a study showing higher levels of omega 3 fatty acids in organic milk, as against ordinary milk.[65] Professor Marion Nestle has gone on the record saying that "I don't think there is any question that as more research is done, it is going to become increasingly apparent that organic food is healthier."[66]

Organic food is not without its problems, however. One is that the rise of organic food, almost always at a higher cost than conventionally farmed food, has perpetuated the idea that an unadulterated diet is something that concerns only the rich. This is not the fault of organic farming per se, though it is not helped by the fetishizing of organic tidbits in exorbitant food boutiques. As Derek Cooper has written, there are effectively now "two kinds of food":

> There is the cheap and nasty stuff—wretched sausages made with mechanically recovered gristle and slurry, nonnutritious packets and tins of highly coloured rubbish and snacks rich only in empty calories—and the expensive foods which attract the words *real, natural, organic, traditional, pure, handmade.* But shouldn't all food be as safe and as pure and as fresh as possible? Why have cheap bad food at all?[67]

There is a danger that organic food becomes a rich person's preserve, a means of protecting the privileged from the squalid realities of poverty. More good would be done by reducing pesticide use across the board—something that is starting to happen, encouragingly, in many sectors of agriculture—and switching wherever possible to sustainable and humane farming than by continuing to set "organics" apart from the rest of agriculture.

"Organic" has become a brand like any other, and as with all brands, it has the potential to mislead. It has become a label in which some food shoppers invest all their hopes, viewing it as a guarantee that the food in question is morally spotless, gastronomically perfect, and entirely, reassuringly safe. But there is no such thing as perfect food, and the drive to find it can make people slightly unhinged. A new condition has been diagnosed in affluent modern societies: orthorexia, or a fixation with righteous eating.[68] Unlike anorexics, orthorexics do not particularly want to get thin; they only want to eat the healthiest, most ecologically sound food possible. The drive to do so, however, may push them to cut out first one food group and then another, until they are subsisting on an extremely limited and socially isolating diet.[69] One former orthorexic described how his obsession with purity reached a point where it was not enough to have a diet consisting almost entirely of organic vegetables; they had to be organic vegetables that had not been out of the ground for more than fifteen minutes.[70]

Even among those of us who are not technically orthorexic, organic food can seem to mean "food that won't poison me or my children if I throw enough money at the problem." As with Hassall's Pure Food Company, there is a tacit encouragement to abandon your senses and put all your faith in the purity of the brand. But as we have seen time and time again, abandoning your senses is about the worst thing you can do if you don't want to be swindled. Much excellent food is organic; but it doesn't follow that all organic food is excellent. There is plenty of organic food that just doesn't taste very good by comparison with its conventional alternatives. There is also plenty of food made with real integrity and taste that is not technically "organic": is a perfect jar of Scottish heather honey, unctuous and jellylike in its sweet richness, rendered inedible by the fact that it is not officially "organic"?

A more serious problem is that the organic standard can mean many different things. In Britain, organic food certified by the Soil Association has to meet stringent requirements for crop rotation, the kind of manure used to enrich the soil, the way chickens and pigs are housed, fed, and looked after, and so on. Organic food without the Soil Association logo may be less exactingly produced. If you buy an organic chicken or organic eggs on the understanding that the animals who produced them led happy free-ranging lives, you may be sorely mistaken. In America, animal rights activist Peter Singer uncovered eggs that were organic (because they had been fed organic feed) that came from hens leading only marginally less crowded and confined lives than conventional battery birds. Food writer Nina Planck, the leading advocate of farmer's markets, has complained that the U.S. rules on organic produce do not place enough emphasis on putting animals out to pasture: "it's quite possible that the organic bacon or turkey burgers in your refrigerator came from animals that never left the barn."[71]

There are fears that the organic sector is becoming a victim of its own success. Now that the big food companies have got in on the organic act—wanting a slice of the more than twelve-billion-dollar market—they have pushed to dilute the standards, lobbying to include some synthetic chemicals under the permitted definition of organic. Rising demand for organic food in Britain means that more and more if it is air-freighted in from abroad, lengthening the chain between consumer and producer and giving the lie to the ideal of organic food as wholesome and environmentally sound. Given these "food miles," many food campaigners now believe it is better to buy local food—even if not technically "organic"—than it is to buy "organic" vegetables from halfway across the world. Closer to home, certain products—notably organic farmed salmon—have, in the view of many old-school organic farmers, made a mockery of what the organic standard was supposed to stand for.

Meanwhile, some food sold as organic or "free range" has been found to be entirely bogus. Like every other premium food in history, the organic market offers an irresistible temptation to swindlers. A December 2006 brought shaming news for Julie's Restaurant, a celebrity hang-out in Holland Park, one of the posher bits of London, known to have been visited by Gwyneth Paltrow and Kate Moss, among others. The restaurant's menu had made great play of its organic meat: gourmet sausages, spice-crusted rack of lamb, and marinated organic chicken. But officers conducting a routine hygiene inspection found that not a single piece of organic meat had been delivered to the restaurant over a fifty-two-day period, saving the owners approximately £4,200. When the well-heeled customers ordered their dish of organic chicken, they were really being given a plateful of broiler. The restaurant's managing partner, Johnny Eckerperigan, pleaded guilty, saying "it was purely a mistake." He was fined £7,500.[72] Similarly, in May 2006, a butcher in New Zealand was fined $10,000 (NZ) for selling meat as "certified organic" when it was not.[73] There are probably far more scams of this kind than ever make it to court.

It is not just organic food that is susceptible to these fiddles, but "ethical" and "free range" food in general. In November 2006, British police were investigating claims that up to thirty million eggs had been illegally passed off as free range.[74] Shoppers had been duped into paying double the normal price for caged eggs on the under-standing that they were high-quality free-range eggs. Several egg farmers near Coventry were arrested on a conspiracy to defraud. It was a case not just of cheating but potentially of poisoning, since the fraudsters seem to have imported eggs from Spain—where one in eight boxes contains salmonella—and stamped them with the Lion Mark, a guarantee that eggs are salmonella-free. Once again, the fragility of the trust between consumers and producers was exposed; again, here was proof that labels alone are no guarantee of food quality.

Free-range organic eggs. These Waitrose eggs come from hens kept in free-range conditions, fed an organic diet. But there have been instances of ordinary eggs from caged hens being fraudulently sold as "free range."

It is worth putting this free range fraud into perspective, though. However dishonest and low grade the eggs sold by these British swindlers, they are as nothing compared to the worst food piracy going on in modern China and Bangladesh. While Western consum-ers fret about the long-range dangers of carcinogens or the horrors of ingesting some nonorganic chicken, shoppers in the Far East and South East Asia are still often condemned to eat foods which are catastrophically falsified and poisoned, in ways which are not in the least hypothetical.

Fake Eggs and Poisoned Babies

It has been called the world's most "unbelievable" invention. Certainly it is the oddest thing, by far, that I have come across while writing this book, as well as the most cruelly ingenious, if true. In 2005–6, it was widely reported that crooked entrepreneurs in China had devised a way of making entirely fake eggs, from scratch. On the outside, they looked just like ordinary eggs. On the inside, they were a medley of peculiar chemicals that can make those who eat them extremely ill. In 2005, an undercover reporter from *East Week*, a magazine based in Hong Kong, claimed to have succeeded in enrolling on a three-day course on how to make the counterfeit eggs.

The instructor, a young woman, taught the reporter how to construct each separate component of the eggs. First, she mixed together gelatine, benzoic acid, coagulating material, alum, and a further unknown powder to make the "white." For the yolk, some lemon-yellow food colouring and possibly some seaweed was mixed with so-called magic water, a substance containing calcium chloride, which gives the yolk a thin outer membrane. Yolk and white were then shaped in special moulds—a round mould for the yolk and an oval for the white. Finally, a liquid containing paraffin wax was poured over the whole thing and left to set into a hard white shell. When cracked open, these human-made eggs could be cooked just like the real thing, though the shells were more fragile. When fried, they looked just like hen's eggs, except that the white came out bubblier. *East Week* magazine reported that those who had tasted the fake eggs said they tasted similar to real ones. Their effects were not the same, though: the fakes contained no nutrients and apparently could cause stomachache, memory loss, and delirium.

The whole thing defies belief. If it weren't so wicked, you could almost admire the audacity of the fraud, the futile surrealist gesture of *constructing* an egg. As an economic cheat, though, it is baffling. How could a process so laborious conceivably pay off, especially for a product as modest as an egg? As one Chinese blogger asked, "You have to wonder, was the profit margin of manufacturing fake eggs really that much greater than opening a chicken farm and selling the

313

real article?"[75] Well, yes, apparently. The wholesale price of the fake eggs was 0.15 yuan a piece (equivalent to around 3 U.S. cents), half the price of a real egg. Given the low margins in chicken farming, there was a high incentive for these grotesque fabrications (just as, in Accum's day, there was economic incentive to go to absurd lengths to fabricate entirely fake tea). If the entrepreneurs could sell the fake eggs in sufficient volume, they could make a lot of money. The fakes seem to have been smuggled in high volume from China to Vietnam, where they were hard to detect at retail markets in Hanoi and Ho Chi Minh City. In 2005, Dr. Tran Dang of Vietnam's Food Safety and Hygiene Department announced its concerns. "There is no doubt that the fakes are dangerous to your health," he commented.[76]

Since then, doubt has been cast on the whole story. An online academic article detailing the fraud—in the *Internet Journal of Toxicology*—has been taken down without explanation. An editor at *Than Nien News*, a Vietnamese newspaper that originally ran the story, tells me that they now believe the information was "wrong."[77] Perhaps the whole thing was a hoax. In this age of endless information, it is so hard to know what to believe. Either way, the fact that the fake-eggs story gained such wide credence is itself a symptom of the depths that adulteration has now reached in China.

Fake eggs would simply represent the crowning monstrosity in a Chinese economy teeming with fakes. In 2006, there was a thousand-fold increase in the counterfeiting activities of Chinese businesses, according to Andes Lam, a strategic risk assessor.[78] It is not just a question of pirated DVDs, bogus designer handbags, and fake electronic goods, though all of these are rife. Like Germany in the First World War, the big cities of modern China are places where ersatz is a way of life and where many people have come to *expect* the food on sale in markets to be phoney. Food fakes have become so normalized that swindlers now try to get away with ever bolder tricks, such as the "fried tofu cakes" sold in Shanghai, which turned out to be gypsum, paint, and starch, fried in an "oil" made from swill and intestines; or a recent case from December 2006 of a factory manager arrested for making "edible lard" from recycled sewage and industrial oil.[79] In 2004, Zheng Xiaoyu, the head of China's National Food and Medicine

Inspection Bureau (equivalent to the FDA or the FSA; a body set up in China in 1998), launched a "Fear-Free Food Campaign." On national state television, Zheng commented that "It's hard to know what you can eat any more. I have exactly the same food-safety fears as ordinary citizens"—a statement that did not entirely instil confidence, since Zheng himself was the man responsible for making sure that the food eaten in China was safe. A year later, he was dismissed, on suspicion of having taken bribes worth around $850,000 from drugs companies, in exchange for allowing them to evade the drug approval standards.[80] In 2007, Zheng was sentenced to death.[81] Food and drugs in China are still the opposite of fear-free.

A malnourished Chinese baby and its mother. The baby had been fed on fake milk.

The head of the Chinese Academy of Environmental Law, Cai Shouqiu, puts the problem in blunt terms: "Making and selling food is so lucrative and so rampant that we don't have the means to control it. Everyone just wants to make money."[82] Even if that means poisoning babies. Accum wrote in 1820, of the swindlers of London, that the "possible sacrifice of even a fellow creature's life is a secondary consideration" to making money.[83] *Plus ça change.* In April 2004, it came to light that at least thirteen babies died in central China and hundreds more became dangerously ill after being fed fake formula milk.[84] In Anhui province, where the scandal broke, they called it "big head disease." Parents couldn't understand why their babies' heads were ballooning, while the rest of them got thinner. Some even told themselves that the bloated face was a sign of good health.[85] The true answer was that the babies were suffering from the worst malnourishment that doctors had seen in twenty years, thanks to drinking "milk" that really mainly consisted of sugar and starch, with only a tiny percentage of the necessary protein and other nutrients.

Terrifyingly, the fake milk was not the product of just one renegade swindler, but constituted an entire mini industry. An initial state inquiry revealed that forty-five different types of substandard powder were on sale in Anhui, manufactured by more than 141 factories across China.[86] The fake brands all sold at a cheaper price than the premium infant formula sold by Nestlé. Given the deprivation of the Anhui region—where the average rural income is around a dollar a day—low-cost milk was the only kind that most families could afford. It came at a terrible price. Effectively, the fraudsters were profiting from the poverty of their customers.

A journalist from the *New York Times* spoke to Zhang Linwei, a father who in 2003 was earning around sixty dollars a month making bricks.[87] When his wife, Liu Li, gave birth to a daughter, she was unable to produce enough breast milk to feed her baby. Mr. Zhang decided, understandably, to buy a low-cost infant formula recommended by a friend. Even this low-cost brand took eleven dollars a month out of the family budget—more than a sixth of the total income. Mr. Zhang's daughter got through a container of the milk every two or three days, but she put on no weight. At five months, she died. The doctors in the hospital where she was treated said her tiny body was so underdeveloped they could not find a single usable vein for a transfusion. They told her parents that the baby's death was the result of drinking fake milk. Liu Li collapsed in shock, having had no idea that the formula she lovingly fed her daughter had been bad. "These babies are really innocent," said a heartbroken Mr. Zhang. "Whatever you give them to eat and drink, they will take it."

After the baby milk scandal broke, the Chinese government promised stern action. More than 100,000 bags of the fake formula were seized. The premier, Wen Jiabao, announced a nationwide investigation. There were multiple arrests. In Anhui province, forty-seven suspects were detained and paraded in yellow waistcoats at a public shaming ceremony; of these, at least forty were formally charged. The authorities promised they would be dealt with "according to the

Chinese police searching a supermarket for fake baby milk.

law." The first person to be tried was Li Xindao, a shopkeeper in Anhui who had sold some of the fake milk; he was told he should have known the milk was fake because of the low price, and he was sentenced to eight years in prison, plus a fine of 1,000 yuan.[88] Ninety-seven local Communist party officials were also punished for "not discovering or fully investigating" the case of the fake milk.

However harsh these retributions, though, they could not disguise the fact that the fake milk scandal was a deep failure of politics. This is the sort of fraud that flourishes when an unbridled market economy of food coincides with the wrong kind of government. It happened with the laissez-faire government of industrial Britain in the 1820s; it happened in New York in the 1860s, when corrupt Tammany Hall officials allowed swill milk to be produced under their very noses; and it is happening now in twenty-first-century China, where the state only effectively intervenes to stop food piracy only when babies start dying. One British journalist in China has written that the problem of counterfeited food is itself "a consequence of China's economic policy, which has encouraged local provinces to pursue growth at all costs."[89] A blind eye will be turned to pirateers, so long as they pay their taxes and generate profits. A professor at People's University, Huang Guoxiong, was quoted as saying that "there is an outdated belief among local officials that they can only kickstart development in their areas by fostering low-price industries producing fakes."[90] However much the Chinese government lamented the dead babies, it was partly responsible for their deaths. The scandal brought to the surface the feeble regulatory powers of the Food and Medicine Inspection Bureau. In March 2003, state news media reported that of 106,000 food companies in China, only 17,900 were licensed. A survey of the nation's food from 2003 found that almost a fifth of food products did not meet national health standards.[91] In this context, horrors such as the fake milk affair are only to be expected. And so it goes on. The latest scandal, in 2006, was lard made from recycled sewage.

A similar situation prevails in Bangladesh, which enjoys the unenviable reputation of having the most adulterated food in Southeast Asia, and therefore, probably, the world. A conference at Kathmandu, Nepal, in March 2004 compared the adulteration rates of different Asian countries. According to a study conducted by two NGOs, India had a rate of 10 percent, Nepal 15 to 18 percent, and Sri Lanka a more worrying 20 to 30 percent. Bangladesh easily beat them all, however, with an adulteration rate of 45 to 50 percent.

Since swindling is a covert activity, such figures are by their very nature almost impossible to verify; but there are plenty of indications that Bangladesh does suffer an unusually adulterated food supply, thanks to an ineffective legal framework and administration for dealing with it. In 2002, the Institute of Public Health, which analyses foods for sale, found that out of 426 samples of sweetmeats, 423 were adulterated, as were 28 out of 33 samples of ghee, 19 out of 19 samples of butter oil, and 8 out of 8 samples of condensed milk.[92] Judging from this, the estimate of 45 to 50 percent adulteration in Bangladesh starts to look rather low.

In 2003, a Dhaka newspaper reported that "an organized gang of adulterators" was carrying out its business "with impunity" in the metropolis and elsewhere in the country.[93] The Dhaka City Corporation (DCC), which delegated a committee to police the situation, was not unaware of the problem but lacked the tools to deal with it. During the first ten months of 2004, the DCC collected 700 samples of food from the open market that were suspected to have been adulterated. During the same period, 650 shops and other commercial establishments were prosecuted with selling unfit or adulterated food. The trouble was that once found guilty of adulteration, under the lax and antiquated Bangladeshi food law of 1959, a person had the choice of three months in jail or a fine of Tk 200, the equivalent of a paltry 3 U.S. dollars. Any self-respecting swindler would choose the latter, freeing him up to go back to his nefarious activities.

In 2004, an official at DCC stated: "We have almost stopped filing cases against food adulterators. Now we are trying to motivate people not to buy adulterated food." This is easier said than done when, often, there is no alternative to the adulterated produce. From brick dust in chilli powder to illegal fertilizers in rice, from yoghurt watered down with contaminated water to rotten coconuts sold as fresh, from bread made with burnt lubricant to sweetmeats made with toxic colouring intended for use as fabric dye, adulteration in Bangladesh is everywhere. It is hard, too, to warn people off adulterated food when so much of it is designed to look more attractive. Unscrupulous farmers have used vast quantities of toxic hormones and chemicals artificially to ripen green fruits such as mangoes, melons, papayas, guavas, and bananas. To the shopper, this fruit looks unusually luscious. To eat it, though, may be to suffer kidney and liver problems. In the long term, it is carcinogenic. In 2004, the DCC filed twenty-two cases against criminals who had mixed carbide powder—one of the ripeners—with fruit. Needless to say, all of the twenty-two swindlers took the option of paying their three-dollar fine and going free.

In theory, at least, the situation in Bangladesh has since improved. In 2005, politicians finally responded to the shocking state of the nation's food supply. Magistrates in special mobile courts were empowered to prosecute offenders. Between May and September these mobile courts (321 of them) succeeded in lodging 2,885 cases against adulterators. Then in September, the Bangladesh parliament passed an amendment to the Pure Food Act, setting up a food safety advisory body and giving enforcers more power. Under the new law, adulteration could be punished with six months to three years in jail and a hefty fine of 5,000 to 50,000 Tk. Repeated offenders would be required to forfeit their premises and equipment. At last, here was a food law with the teeth to deter the crime.

Initial optimism has already faded, however. In February 2006, a journalist on the *Daily Star* expressed scepticism about the new food safety amendment: "There is doubt about whether these tough provisions will succeed in improving the situation because no matter how sophisticated the laws are, proper implementation is always absent in Bangladesh."[94] Sure enough, insiders serving on the mobile courts reported that the courts failed to function properly much of the time because team members simply could not coordinate themselves. Meanwhile, despite the threat of greater punishments, the gangs of adulterators continued their activities. In July 2006, the *Bangladesh Observer* reported the widespread sale of adulterated produce in the markets of the Rajbari district: edible oil mixed with machine oil, rotten bread mixed into biscuit dough, toxic dyes in ice creams.[95] Most strikingly of all, many of these adulterations happened publicly, without much sense of shame. Local administrators were reported as shirking their duties, failing to do anything to stop it.

At the time of writing, Bangladesh would seem to have all the components that, as we have seen in the course of this book, encourage swindling to flourish: long chains between producer and consumer coupled with a lack of mutual trust, incoherent food laws, a wild undisciplined market economy, a politics that is by turns apathetic and corrupt, and a culture in which consumers feel powerless to complain. One commentator has attributed the endemic adulteration in the country to the complete lack of "consumer rights" there: people feel they have no redress against dandruff shampoo which makes their hair fall out, CDs with half the tracks missing, and mouldy halva.[96] "Violation of consumer rights and lack of business ethics has become such a common practice that even people with education are failing to protect their rights." And, in a vicious circle, because people do not complain at the shoddy way they are treated, they continue to be treated shoddily.

Whether Bangladesh is condemned to carry on eating the most adulterated food in Southeast Asia is another matter. This is not the 1820s. Bangladesh has one great asset on its side, which is belonging to an information age that can transmit knowledge with greater speed than ever in the past. Though its food is much more falsified than that of the West, its journalism is of just as high a standard as that of Europe and America. While the swindlers operate in darkness, journalists have exposed their activities online with lightning speed, shaming both swindlers and politicians. This welter of information enables the world to learn about Bangladesh; it also gives Bangladesh access to the latest information on ways to combat food fraud. Will information eventually win out against the forces of deception? You would hope so. But history suggests that all victories in this struggle are the presage of further battles to come.

Epilogue

ADULTERATION IN THE TWENTY-FIRST CENTURY

If some humans adulterate food, other humans can stop them.
                    —The London Food Commission (1988)

Adulteration, like poverty, seems to be always with us. The motive to swindle—greed—is a constant in human history. On the other hand, both the incentives and the opportunities for swindling have varied hugely over time. The incentives are mainly determined by economics, and the opportunities by politics and science. Swindling is not created by abstract entities such as "free trade" or "globalization." It is caused by the working of these and other forces on individual human beings, who have other qualities of character than the mere impulse to defraud their fellow citizens. It is not too far-fetched to imagine a society where the right economic policies and the right politics could more or less eliminate food swindling, just as Jeffrey Sachs has argued that an "end to poverty" is not impossible given the political will.[1]

What, though, *is* the right way of fighting the swindlers? In the course of this story, we have seen plenty of alternatives, and none of them is perfect or sufficient in itself. The most basic response of all is to remove oneself altogether from the causes of swindling and retire to a pastoral idyll of some kind or other and attempt to eat what Sylvester Graham called an "Edenic" diet. If adulteration is caused by industrialization, then why not act as though the industrial revolution had never happened, and lead a life of quiet simplicity and self-sufficiency in the country, where you can trust all your food to be good because you grew or raised it yourself? For individuals fortunate enough to do it, this undoubtedly works. If you live like this, good luck to you. Elements of this idyll are undoubtedly useful, too, in the fight against adulteration in general. The excessive use of pesticides in modern agriculture has been rightly lobbied against by organic campaigners; and as a result, chemical pesticide limits are controlled more tightly than they were before. The ideal of self-sufficiency is also a useful corrective to affluent societies in which consumers often feel powerless to control what they eat. As applied to an entire society, however, the pastoral idyll is both utopian and retrograde. For better or worse, most of us live in mass commercial societies. A solution that depends on removal from this reality is no solution at all.

Most food legislation from 1860 onwards has adopted a more modern and realistic approach: fighting the swindlers under the great banner of Safety. The virtue of this strategy has been that it addresses the worst consequences of adulteration. A society with robust food safety laws, and expert officials to administer them, is not going to suffer the horrors of present-day Bangladesh. On the other hand, an excessive preoccupation with safety can make good food worse, as well as leading to unnecessary panics. As we saw in the last chapter, the filth clause in U.S. food legislation has led to overuse of pesticides. Equally, we are all familiar with stories of delicious unpasteurized cheeses being condemned as a threat to "health and safety." And "food safety" can also neglect the moral dimension in adulteration: while strong on poisoning, it is often weak on cheating. There is plenty of fake and substitute food that is technically safe to eat—at least so far as we are aware at the present time—but that does not make it good or honest food, nor does it mean that the people supplying it are not trying to defraud us.

Therefore, many campaigners, as we have seen, have latched on to the ideal of Purity rather than mere Safety. Arthur Hill Hassall in Britain and Harvey Washington Wiley in the United States held up "Pure Food" as the model to aim for. This enabled them to achieve many great things. By setting the bar so high, they were able to restore some integrity to the process of buying and selling food. Hassall and

Wiley addressed cheating as well as poisoning, adopting a zero-tolerance attitude toward trickery. It was not acceptable to sell "pure mustard" that was not pure, or "honey" that was really glucose, or "sweet corn" that was sweet only on account of the saccharine that had been added. The down side of "Pure Food," though, is that there is ultimately no such thing; both Hassall and Wiley ended up in futile quests for absolute nutritional purity, which set them ever further apart from the ordinary consumer. The promise of Pure Food can ultimately lead to a disengagement of the senses. Consumers are encouraged to surrender their judgement over whether food is good or bad to experts. In the process, the simple pleasures of good food may be forgotten.

A more consumer-led approach has been to counter adulteration with information in the form of publicity and labelling. The press has always played a role in exposing food fraud, which has been heightened at various times through the work of muckrakers, from Upton Sinclair to Ralph Nader. In addition, we now have whole dictionaries of information about food at our disposal in the form of labelling. The right kind of publicity has been extremely effective in curbing the swindlers. After *The Jungle*, the meatpackers did clean up their act (though not to the extent that Sinclair wished). Mandatory food information on labels has eliminated whole swathes of swindling. Publicity is a double-edged sword, though. If there is muckraking, there is also advertising. Moreover, as we have seen, there is a fine line between useful science-based exposure of fraud and scaremongering. When swindling is exposed in the wrong way, it can lead not to firm action but to apathy or hysteria, neither of which helps us to eat better. By the same token, trusting too much to the label can be counterproductive. In and of itself, labelling is not enough to enable people to choose food wisely and well.

For this, we need knowledge of food—real trustworthy knowledge, rather than empty information. Of all the approaches to fighting swindling we have seen, the best has been that which fights fake food with a sound appreciation of the real thing, because in most cases this gives the swindler nowhere to hide. In some ways, this is a very old approach—we saw it in the medieval guilds, and their modern-day equivalents, the AOC and PDO systems, which start with a knowledge of what a particular food should be like and then do everything they can to protect it. But real knowledge of food is not backward-looking, because it deepens as our knowledge of science deepens.

Another version of this approach has been adopted by cutting-edge scientists who use the latest scientific techniques to expose food fraud—whether Accum with his chemistry sets in 1820 or Mark Woolfe with his DNA tests today. However sophisticated their science, Accum and Woolfe both started from a simple enjoyment of food, and a desire to stop people ruining that pleasure. There are many virtues in fighting adulteration with knowledge of good food. It works not just for agrarian societies but for the modern industrial democracies that most of us now live in. It reengages the consumer with food, thus counteracting the problem of the long chain between producer and eater. Unlike so many of the other ways of fighting the swindlers, it does not stifle pleasure, or unduly heighten fear. It lessens the incentive to swindle, since the more people understand the difference between what is fake and what is real, the harder it will be to pass one off as the other, and to make money out of the difference.

The only significant problem I can see with fighting adulteration with real knowledge about food is that many of us don't have it. If a whole society was serious about eliminating food fraud, it would embark on a vast overhaul of the education system, making cooking and a wide practical knowledge of food an essential part of learning for all ages. We would need to be taught what most medieval eaters knew instinctively: what bread tastes like when it is made from nothing but flour, water, salt, and leavening; what ham tastes like when it hasn't been injected with excess water; how Basmati rice smells and how it differs from standard long-grain rice. Children in schools should be taught how vegetables are grown—without pesticides—and how they can be cooked—without additives; and how different kinds of apples taste; in the kind of classroom activities pioneered in Berkeley by Alice Waters's Edible Schoolyard. Consumers generally would need to be taught the difference between the fake and the real: between the packaged pineapple pie of *100 000 000 Guinea Pigs* and a real one made from fresh fruit, butter and flour. It would take a brave government to embark on teaching consumers such things; for however much we are lectured about fat, sugar, and obesity, few governments (if any) have ever wanted to teach us about food to the extent of antagonising the food industry, as Marion Nestle's work has shown.[2] We would also need to be educated about the true cost of food, so that the next time we are presented with a luxury food that is implausibly cheap, we know enough to question whether all is as it seems.

This would be a huge undertaking, but even so it would not be enough on its own, because so many frauds are invisible. Government would also have to make sure it kept food laws and enforcement up-to-date, on the basis of the latest scientific knowledge, in order to keep one step ahead of the fraudsters (or at least, to make sure the fraudsters aren't always one step ahead of the people trying to stop them). Much of this science—from DNA testing to mass spectroscopy—is bound to be too technical for the general public to understand, remaining the preserve of experts. We as citizens would therefore have to trust these scientists to do the right thing. With the confidence of the latest science behind them, governments would have to be fearless about prosecuting swindlers, when they are found out. The story of adulteration has been a story of the repeated failure of modern politics to value consumer interests above those of the market. The future could be different, though. Bad and falsified food is not a necessary fate for modern societies. With climate change, food is becoming more expensive and contested after the first period in Western history of relative abundance. In a world preoccupied with waste, there is every opportunity and reason now for governments to take more responsibility for the food supply—just as they are taking more responsibility for clean air and carbon emissions—not by telling us what to eat but by eliminating poisonous and deceptive foods from production.

For the moment, though, most of this is yet to happen. In the history of food adulteration, government intervention to stop bad food has always come later than it should; and it has never been adequate to the problem. While we are waiting—and where possible, lobbying—for governments to do more to improve the food supply, we as individual consumers should do what is in our power to prevent ourselves and our families being cheated and poisoned. If we embrace even a simple understanding of what makes good food, there are plenty of ways we can minimize, if not eradicate, the risks of being taken for a ride.

If you don't want to be swindled, here is what you can do (assuming you are fortunate enough to live in a country where basic food standards are enforced). It won't be enough, but it will be a start. Buy food fresh, in whole form. Buy organic, where possible. Buy your food from someone you trust; if they live nearby, so much the better. Cook it yourself and familiarize yourself with the ingredients that go into proper food, so that when you are served a fake you will know the difference, and have the confidence to complain. Above all, trust your own senses. You know more than you think you do. Hear the noise a piece of good chocolate makes when it snaps; see the gleam on a fish that is truly fresh; taste the sweetness in a fresh cinnamon stick; inhale the perfume of real Basmati rice. Wake up and smell the coffee.

Notes

Full details of the works cited in short form can be found in the bibliography.

Preface

    1. McGee (1984), p. 536.

    2. Barton Hutt (1978), p. 507.

1: German Ham and English Pickles

    1. See Accum, *Culinary Chemistry* (1821), passim.

    2. Accum, *A Treatise on Adulterations of Food* (1820), p. 13.

    3. Reproduced in Accum, *A Treatise on the Art of Making Wine* (1820), p. x.

    4. Ibid., p. xxii.

    5. Accum (1966), p. 32.

    6. Accum, *Treatise on Adulterations*, (1820), p. 31.

    7. Ibid., p. 32.

    8. Ibid., p. iii.

    9. Smollett, *Humphry Clinker* (1771), chap. 38.

    10. Reproduced in Accum, *Treatise on Wine*, p. xxi.

    11. Browne (1925).

    12. Accum (1815), p. 77.

    13. On Winsor, see Everard (1949), pp. 17–26, and Williams (2004), passim.

    14. Accum (1815), p. 187.

    15. Ibid., pp. 170–73.

    16. Browne (1925), p. 829.

    17. See, for example, ibid., p. 832; Cole (1951), p. 128.

    18. Accum, (1821), Bread, pp. 22–23.

    19. Rumohr (1993), p. 126.

    20. Ibid., p. 116.

    21. Hughson (1817), pp. 196–97.

22. Stieb (1966), p. 163.
23. Accum, *Accum's Chemical Amusements* (1817), p. iii.
24. *European Magazine*, June 1820.
25. Browne (1925), p. 839.
26. Ibid., p. 839, reminiscence of Silliman.
27. Browne (1925), p. 845.
28. Ibid., p. 847.
29. Ibid., pp. 846–47.
30. Hudson (1992), p. 61.
31. Accum (1966), p. 21.
32. Filby (1934), p. 18.
33. Accum (1966), pp. 222–23.
34. *Philosophical Magazine*, vol. 54 (1819), p. 218.
35. Accum (1966), p. 239.
36. Ibid., p. 244.
37. Accum, Bread, p. 128.
38. Quoted in Browne (1925), p. 1031.
39. Accum, *Treatise on Adulterations* (1820), p. 14.
49. Accum, *Culinary Chemistry* (1821), p. 59.
41. Accum (1966), pp. 25–28.
42. Spencer (2002), pp. 208, 246.
43. Acton (1993), p. 21.
44. Accum, *Culinary Chemistry* (1821), p. 55.
45. Ibid., pp. 249–56.
46. Accum, *Treatise on Wines* (1820), pp. 48, 46.
47. Accum, *Culinary Chemistry* (1821), p. 150.
48. Ibid., p. 309.
49. Accum, *Treatise on Adulterations* (1820), pp. 244.
50. Rundell (1818), p. 283.
51. Accum (1966), p. 185.
52. Accum, *Culinary Chemistry* (1821), p. 315.
53. Ibid., p. 27.
54. Ibid., p. 24.
55. Accum (1966), p. 98.
56. Ibid., p. 100.
57. Accum, *Culinary Chemistry* (1821), p. 4.
58. Accum (1966), p. 218.
59. Ibid.
60. Ibid., p. 225.
61. Accum, *Culinary Chemistry* (1821), p. 331.
62. Accum (1966), pp. 231, 233.
63. Accum, *Culinary Chemistry* (1821), p. 332.

64. Accum (1966), pp. 232–33.
65. Ibid., p. 31.
66. Ibid., p. 224.
67. Ibid.
68. Letter to *The Times*, 29 May 1824.
69. Accum (1966), p. 19.
70. Ibid., pp. 206–10.
71. Ibid., p. 210.
72. Ibid., p. 15.
73. Ibid., p. 20.
74. Ibid., p. 16.
75. Ibid., pp. 211–14.
76. Ibid., pp. 163–70.
77. Ibid., p. 172.
78. Ibid., p. 22.
79. Ibid., p. 23.
80. Ibid., p. 22.
81. Ibid., p. 23.
82. *The Times*, 5 March 1818, p. 3.
83. Accum (1966), pp. 126, 116.
84. Ibid., p. 157.
85. Child (1798), pp. 6, 21.
86. Accum (1966), pp. 143–44.
87. Monckton (1966), p. 159.
88. Accum (1966), p. 123.
89. Patton (1989).
90. Spencer (2002), p. 263.
91. Accum (1966), p. 148.
92. Taylor (1972), pp. 50–51.
93. For accounts of the book mutilation affair, see Browne (1925); Cole (1951); Gee (2004).
94. Greenaway (1971); *Archives of the Royal Institution of Great Britain*, vol. 6, 16 April 1821.
95. Quoted in Browne (1925), p. 1142.
96. "Mr Frederick Accum," letter to *The Times*, 10 January 1821, signed "A.C."
97. "Mr Accum's Case," *The Times*, 6 April 1821.
98. Reynolds (1822), note to line 161.
99. Browne (1925), p. 1140.
100. Cited Cole (1951), p. 141.
101. Greenaway (1971), vol. 6, 23 December 1820.
102. Parmentier (1803), p. 181.

2: A Jug of Wine, a Loaf of Bread

1. Phillips (2001), p. 41.
2. Phillips (2000), p. 32.
3. Pliny (1968), book 14: 130, vol. 4, p. 273.
4. Ibid., book 14: 17, vol. 4, p. 197.
5. Quoted in Cato (1933), pp. 48–49.
6. Juvenal (1984), satire 5.
7. Phillips (2000).
8. Pliny, book 23: 45–46.
9. Columella, *De Re Rustica*, 12: 19–21.
10. Cato (1933), pp. 48–49.
11. Phillips (2000), p. 34.
12. Eisinger (1982), p. 298.
13. Ibid., p. 294.
14. Filby (1934), p. 140.
15. Ibid., p. 146.
16. Accum, (1966), p. 82.
17. Filby (1934), p. 145.
18. Drummond and Wilbraham (1939), p. 47.
19. Addison and Steele, *The Tatler* vol. 2 (1797), p. 110.
20. Loubère (1978), p. 73.
21. Ibid., p. 166.
22. Quoted Phillips (2000), p. 32.
23. Fielden (1989), p. 21.
24. Filby (1934), p. 131.
25. Eisinger (1982).
26. Fielden (1989), p. 4.
27. Ibid., p. 165.
28. Ibid., p. 6.
29. Filby (1934), p. 130.
30. Fielden (1989), p. 19.
31. Ibid., pp. 5–6.
32. Accum (1966), p. 79.
33. Filby (1934), p. 158.
34. Dillon (2004), passim.
35. Drummond and Wilbraham (1939), p. 47.
36. Redding (1833).
37. Loubère (1978), pp. 252–53.
38. Stanziani (2003), p. 128.
39. Ibid.
40. Loubère (1978), p. 166.

41. Stanziani (2003), p. 137.
42. Robinson (1999), p. 4.
43. Fielden (1989).
44. Atkin and Lee (2005).
45. Catchpole (2006); Shaugnessy (2005).
46. "Austrian Wines," *New York Times*, 22 October 1986.
47. Haydon (2001).
48. Zupko (1977), p. 26.
49. Ibid., p. 27.
50. Ibid., p. 36.
51. Studer (1911), p. xxi.
52. Zupko (1977), p. 36.
53. Studer (1911), p. xxi.
54. Ibid., p. xxvi.
55. Drummond and Wilbraham (1939), p. 40.
56. Studer (1911), p. xxvii.
57. Kaplan (1996), p. 2.
58. Ibid., p. 475.
59. Ibid., p. 471.
60. Ibid., p. 479.
61. McCance and Widdowson (1956), p. 32.
62. Garnsey (1988), pp. 28, 29.
63. Platt (1596), p. 1.
64. Camporesi (1989).
65. Smith and Christian (1984), pp. 347ff.
66. Ibid.
67. Ibid.
68. *Observer*, 15 February 2004.
69. On the long history of fears of food, see Ferrières (2006).
70. Jackson (1758), p. 12.
71. Drummond and Wilbraham (1939), p. 222.
72. McCance and Widdowson (1956), pp. 23–24.
73. Ibid., p. 26.
74. Manning (1757), p. 4.
75. Anon., *Poison Detected* (1757), p. 6.
76. Markham (1757), p. 22.
77. Anon. (1757), pp. 3–4.
78. Ibid., pp. 8, 16.
79. Manning (1757), p. 12.
80. Collins (1758), p. 37.
81. Ibid., p. 19.
82. Drummond and Wilbraham (1939), p. 226.

83. Markham (1757), p. 6.
84. Jackson (1758), p. 8.
85. Filby (1934), p. 99.
86. Ibid., p. 101.
87. Manning (1757), pp. 3–4.
88. Jackson (1758), p. 14.
89. http://www.ilo.org, accessed 31 May 2006.
90. Accum (1966), p. 106.
91. Drummond and Wilbraham (1939), p. 349.
92. David (1994), p. 191.
93. Ibid., p. 193.
94. Smollett (1771), chap. 38.
95. Jackson (1758), p. 13.
96. McCance and Widdowson (1956), p. 32.
97. Anon. (1757), p. 54.
98. Quoted McCance and Widdowson (1956), p. 32.
99. Rubin (2005), p. 135.
100. Renard (1918), p. 34.
101. MacKenney (1987), p. 18.
102. Renard (1918), p. 33.
103. Ibid., p. 38.
104. Ibid., p. 33.
105. Dorey (2007).
106. Swanson (1989), p. 17.
107. Ibid., p. 22.
108. Quoted in Kaplan (1996).
109. Renard (1918), p. 55.
110. Patton (1989), p. 9.
111. Whittet (1968), p. 801.
112. Toussaint-Samat (1992), p. 493.
113. Filby (1934), p. 24.
114. Whittet (1968), p. 803.
115. Filby (1934), p. 25.
116. Shipperbottom (1993), p. 247.
117. Filby (1934), p. 27.
118. Quoted in ibid., p. 30.
119. Shipperbottom (1993), p. 251.
120. Filby (1934), p. 28.

3: Government Mustard

1. Eliot (1884) p. 323.
2. Hassall (1855), p. 160.

3. *Punch*, vol. 20, 1 February 1851, p. 44.
4. Tickletooth (1999), p. 185.
5. Mayhew (1980), vol. 2, pp. 322–23.
6. Ibid., p. 323.
7. Ibid.
8. Freeman (1989), p. 11.
9. Engels (1993), p. 80.
10. Mayhew (1980), vol. 2, p. 2.
11. Ibid.
12. Engels (1993), p. 80.
13. Mayhew (1980), p. 252.
14. Freeman (1989), p. 26.
15. Engels (1993), p. 80.
16. Ibid., p. 81.
17. Tickletooth (1999), p. 185.
18. Mayhew (1980), p. 260.
19. Ibid., p. 252.
20. Ibid., p. 260.
21. Engels (1993), p. 112.
22. Burnett (1989), chap. 5.
23. Anon. (1855b), p. 249.
24. See, for example, Normandy, Chevallier.
25. Acton (1857), p. 1.
26. Chevallier (1854), p. 138.
27. Anon. (1855a), p. 185.
28. Burnett (1989), chap. 5.
29. Ibid.
30. Acton (1857), p. 19; Burnett (1989), chap. 5.
31. Burnett (1989), chap. 5.
32. Anon. (1855a), p. 57.
33. Chevallier (1854), vol. 2, p. 173.
34. Acton (1857), p. 28.
35. Mitchell (1848), p. xi.
36. Ibid., p. x.
37. Acton (1857), p. 17.
38. Anon. (1851), p. 81.
39. Ibid., p. 9.
40. Acton (1857), p. 31.
41. Mitchell (1848), p. x.
42. Stanziani (2005), pp. 51, 52.
43. Anon. (1851), p. 43.
44. Normandy (1850), p. 81.
45. Anon. (1851), p. 43.

46. Mitchell (1848), p. 155.
47. Anon. (1851), p. 40.
48. Ibid., p. 45.
49. Normandy (1850), p. 79.
50. Chevallier (1854), p. 138.
51. Anon. (1851), p. ix.
52. Anon. (1830), pp. 135, 127.
53. Ibid., pp. 117–18.
54. Ibid., p. 33.
55. Quoted in Drummond and Wilbraham (1939), p. 345.
56. Letheby (1870), p. 265.
57. Mitchell (1848), pp. 79, 186.
58. Ibid., pp. vi, vii.
59. Ibid., p. 41.
60. Stieb (1966), pp. 52–56.
61. Ibid., p. 28.
62. Cited in Hassall (1893), p. 44.
63. Clayton (1908), p. xiii.
64. Stieb (1966), p. 175.
65. Hassall (1893), p. 47.
66. Ibid., p. 43.
67. Ibid.
68. "Record of the Results of Microscopial and Chemical Analyses of the Solids and Fluids Consumed by All classes of the Public. Coffee and Its Adulterations [Second Report]," *Lancet*, vol. 57, no. 1443, 26 April 1851, p. 466.
69. Hassall (1893), p. 43.
70. Gray (1983), p. 99.
71. Hassall (1893), p. 43.
72. See Rowlinson (1982), p. 64, for illustration.
73. Hassall (1893), p. 44.
74. Rowlinson (1982), p. 65.
75. Stieb (1966), p. 179
76. Hassall (1893), p. 44.
77. Ibid., p. 46.
78. Gray (1983), p. 103.
79. "Spices and Their Adulterations," *Lancet*, vol. 59, no. 1487, 28 February 1852, p. 226.
80. Hassall (1893), pp. 50–51.
81. Gray (1983), p. 106.
82. Hutchins (1909), p. 15.
83. Quoted in Gray (1983), p. 101.

84. "Record of the Results of Microscopial and Chemical Analyses of the Solids and Fluids Consumed by All Classes of the Public: Water and Its Impurities," *Lancet*, vol. 60, no. 1511, 14 August 1852, p. 257.

85. Ibid.

86. Hassall (1893), p. 62.

87. Ibid., p. 69.

88. Ibid., p. 46.

89. "Arrow-Root and Its Adulterations," *Lancet*, vol. 57, no. 1431, 1 February 1851, p. 143.

90. Hassall (1855), p. 175.

91. "Report by the Analytical Sanitary Commission on Farinaceous Foods," *Lancet*, vol. 57, no. 1451, 21 June 1851, pp. 675–79.

92. "Report by the Analytical Sanitary Commission on Farinaceous Foods," *Lancet*, vol. 57, no. 1450, 14 June 1851, p. 654.

93. Nelson (2005).

94. "Report by the Analytical Sanitary Commission on Ervalenta, Revalenta, etc., etc.," *Lancet*, vol. 57, no. 1450, 14 June 1851, p. 657.

95. Ibid., p. 656.

96. Ibid., p. 658.

97. Ibid., p. 659.

98. Anon. (1855a), p. 42.

99. Stieb (1966), p. 105.

100. Rowlinson (1982), p. 66.

101. Burnett (1989).

102. Letheby (1870), p. 273.

103. Anon. (1855a), p. 10.

104. *The Times*, 15 October 1873, letter.

105. Rowlinson (1982), p. 66.

106. Clayton (1908), p. 84.

107. Anon. (1855a), p. 220.

108. Allingham (1884).

109. Letheby (1870), p. 273.

110. "Report on Poisonous Bottled Fruits and Vegetables," *Lancet*, vol. 60, no. 1511, 21 August 1852, p. 135.

111. Anon. (1855a), p. 41.

112. Ibid., p. 125.

113. Rowlinson (1982), p. 71.

114. Ibid., p. 67.

115. Burnett (1989), p. 229.

117. Ibid., chap. 10.

117. Hassall (1893), pp. 124–25.

118. Ibid., pp. 94–95.

119. Ibid., p. 123.
120. Ibid., pp. 125–26.
121. Ibid., p. 126.

4: Pink Margarine and Pure Ketchup

1. Wiley (1930), p. 199.
2. Goodwin (1999), p. 48.
3. Wiley (1930), p. 199.
4. Young (1989), p. 95.
5. Barton Hutt (1978), p. 508.
6. Block (2004).
7. "Distillery Milk," *New York Times*, 18 August 1854.
8. "Our Food and Drink," *New York Times*, 14 January 1869.
9. "Death in the Jug," *New York Times*, 22 January 1853.
10. "Swill Milk and Infant Mortality," *New York Times*, 22 May 1858.
11. Fildes (1986), p. 168.
12. Apple (1987), p. 9.
13. Lee (2006).
14. Brosco (1999), p. 480.
15. Lee (2006), p. 7.
16. *David Copperfield*, chap. 61.
17. Atkins (1991), pp. 320–21; Rowlinson (1982), p. 70.
18. Atkins (1991), pp. 335, 320.
19. "Death in the Jug."
20. "Distillery Milk."
21. Ibid.
22. "Death in the Jug."
23. Ibid.
24. Ibid.
25. "They Ought to Be Beaten," *New York Times*, 28 October 1878.
26. "Suddenly Dropping Dead," *New York Times*, 3 May 1887.
27. "New York City Swill Milk, Meeting of the Committee of the Board of Health," *New York Times*, 1 June 1858.
28. Young (1989), p. 38.
29. "Pure and Impure Milk," *New York Times*, 21 July 1874; "Swill Milk in San Francisco," *New York Times*, 18 November 1886.
30. "The Annual Report of the City Inspector," *New York Times*, 21 January 1861.
31. "How to Secure Pure Milk," *Washington Post*, 16 June 1893.
32. "What Is Adulteration?" *New York Times*, 19 December 1873.

33. Schmid (2006).
34. "Pickled Poisons," *New York Times*, 19 March 1871.
35. "Adulterated Food," *New York Times*, 5 November 1872.
36. "What Is Adulteration?"
37. "Pickled Poisons."
38. "Food Adulteration," *Washington Post*, 9 January 1881.
39. Goodwin (1999), p. 43.
40. "Food Adulteration."
41. Goodwin (1999), p. 137.
42. Ibid., pp. 27, 28.
43. Slotnick (2004), pp. 573–74.
44. Goodwin (1999), p. 65.
45. Ibid., p. 41.
46. Ibid., p. 137.
47. Obituary of George Thorndike Angell, *New York Times*, 17 March 1909.
48. Young (1989), p. 47.
49. Goodwin (1999), p. 73.
50. Nestle (2006).
51. Levenson (2001), p. 174.
52. http://www.cbc.ca/stories/1999/05/26/business/butter, accessed 11 August 2006.
53. Levenson (2001), p. 173.
54. Ibid., p. 174.
55. Young (1989), pp. 82, 66.
56. Ibid., p. 83.
57. "Growth of Oleomargarine," *New York Times*, 18 January 1886.
58. Twain (1996), p. 412.
59. Young (1989), p. 79.
60. "Tax It Out of the Market," *New York Times*, 19 February 1886.
61. Young (1989), p. 75.
62. "A Plea for Honest Butter," *Washington Post*, 27 September 1880.
63. Young (1989), p. 84.
64. Ibid., pp. 86, 84.
65. "Making Oleomargarine Odious," *New York Times*, 20 May 1886.
66. "Counterfeit Butter," *Washington Post*, 4 February 1879.
67. "Oleo as Adulterant," *New York Times*, 23 May 1886.
68. "He Got Oleomargarine Though He Asked for Butter at a Railroad Restaurant," *New York Times*, 21 December 1886.
69. "An Honest Measure." *Washington Post*, 13 June 1894.
70. Young (1989), p. 87.

71. Levenson (2001), p. 171.

72. "Oleomargarine in Minnesota," *New York Times*, 22 December 1897.

73. "Oleomargarine Laws Invalid," *New York Times*, 24 May 1898.

74. "Dr Wiley, Famous Chemist and Fighter for Pure Food," *Boston Globe*, 8 November 1908.

75. Wiley (1930), pp. 40, 41.

76. Ibid., p. 156.

77. "Dr Wiley, Famous Chemist and Fighter for Pure Food."

78. Anderson (1958), p. 20.

79. Wiley (1930), p. 26.

80. Anderson (1958), p. 33.

81. Young (1989), p. 68.

82. Anderson (1958), p. 22.

83. Ibid.

84. "What Is In Your Food," *Washington Post*, 19 April 1897, p. 10.

85. Wiley (1930), p. 150.

86. Ibid.

87. Ibid., p. 54.

88. Gaughan (2004), p. 4.

89. Wiley (1906), p. 1.

90. Coppin and High (1999).

91. Ibid., p. 5.

92. Anderson (1958), p. 127; Wiley (1907), pp. 208–9.

93. "What Is in Your Food."

94. Young (1989), p. 143.

95. "What Is In Your Food."

96. Young (1989), p. 155.

97. "The Army Meat Scandal," *New York Times*, 1 February 1899.

98. Keuchel (1974), p. 252.

99. Ibid., p. 258.

100. Young (1989), pp. 139, 137.

101. Ibid., p. 143; Anderson (1958), p. 130.

102. Young (1989), p. 143.

103 Wiley (1930), p. 215.

104. Anderson (1958), pp. 149–51; Young (1989), pp. 153–54.

105. "Eating by Rigid Rule," *Washington Post*, 22 December 1902.

106. "Borax Ration Scant," *Washington Post*, 23 December 1902.

107. Wiley (1930), p. 217.

108. Wiley (1907), p. 37.

109. "Effects of Using Borax," *Washington Post*, 2 June 1902.

110. "Dr Wiley in Despair—One Boarder Becomes Too Fat and Another Too Lean," *Washington Post*, 16 December 1902.

111. Young (1989), p. 154.

112. "Borax Begins to Tell," *Washington Post*, 26 December 1902.

113. "Borax Ration Scant."

114. "Poison Eaters Glad—Two Month Vacation for Dr Wiley's Boarders," *Washington Post*, 10 June 1903.

115. "Prof. Wiley's Experiments," *Washington Post*, 2 January 1903.

116. Quoted in "Dr Wiley's Experiment," *Washington Post*, 24 December 1902.

117. "Baby Class in Borax." *Washington Post*, 11 January 1903.

118. Quoted in Murphy (2001), part 2.

119. Quoted in Wiley (1930), p. 219.

120. Ibid.

121. Murphy (2001), part 2.

122. Smith (2001), p. 79.

123. Wiley (1930), p. 220.

124. Young (1989), p. 202.

125. Goodwin (1999), p. 160.

126. Wiley (1930), p. 230.

127. Denby (2006), p. 73.

128. Sinclair (1906), chap. 14.

129. Young (1989), p. 222.

130. Sinclair (1963), p. 120.

131. Ibid., p. 134.

132. Suh (1997), p. 92.

133. Sinclair (1906), chap. 7.

134. Ibid., chap. 14.

135. Ibid., chap. 7.

136. Ibid., chap. 31.

137. Young (1989), p. 226.

138. Sinclair (1963), p. 135.

139. Gottesman (1985), p. xxiii.

140. Young (1989), p. 138.

141. "Inquiry into Beef Trust," *New York Times*, 17 April 1902.

142. "Trust Hunt Critics Answered by Moody," *New York Times*, 23 July 1905.

143. *New York Times*, 28 October 1905; Young (1989), p. 227.

144. Young (1989), p. 226.

145. "Biggest of Trusts," *New York Times*, 29 October 1905.

146. Roosevelt (1954), pp. 179, 180, 178.

147. Young (1989), p. 251.

148. Sinclair (1963), p. 129.

149. Young (1989), p. 233.

150. "President Hunts in 'The Jungle,'" *Chicago Daily Tribune*, 10 April 1906.

151. "Find 'The Jungle' 95 Per Cent Lies," *Chicago Daily Tribune*, 11 April 1906.

152. Roosevelt (1954), letter 3881.

153. Sinclair (1963), p. 129.

154. Young (1989), p. 239.

155. "Author of 'The Jungle' Urges President to Publish Report," *Washington Post*, 29 May 1906.

156. "Sinclair Gives Proof of Meat Trust Frauds," *New York Times*, 28 May 1906.

157. Wiley (1907), p. 554.

158. "Meat Inspection Bill Passes the Senate," *Washington Post*, 26 May 1906.

159. "Meat Trust in a Pickle," *Washington Post*, 31 May 1906.

160. Brantz (2006).

161. Fabian Society (1899).

162. Roosevelt (1913), p. 483.

163. Wiley (1907), pp. 525–31.

164. Young (1989), p. 218.

165. Wiley (1907), p. 528.

166. "Label Must Be Exact," *Washington Post*, 17 November 1908.

167. Wiley (1930), p. 204.

168. Young (1989), p. 218.

169. Ibid., p. 260.

170. Ibid., p. 267.

171. "Canners for Pure Food," *New York Times*, 16 February 1907.

172. Kolko (1963), pp. 108–10.

173. Young (1989), p. 264.

174. Smith (2001), p. 85.

175. "Against Benzoate of Soda," *Washington Post*, 30 November 1908; Young (1989), p. 215.

176. Young (1989), p. 216.

177. Wiley (1907), pp. 316–17.

178. Smith (2001), p. 79.

179. Ibid., p. 86.

180. Ibid., pp. 85–89.

181. Potter (1959), p. 69.

182. Ibid.

183. Coppin and High (1999), pp. 121–25.
184. Smith (2001), p. 87.
185. Potter (1959), pp. 67–68.
186. Smith (2001), p. 87.
187. *New York Times*, 22 April 1909.
188. Smith (2001), p. 97.
189. Ibid., p. 110.
190. Ibid., p. 111.
191. Ibid., p. 91.
192. Wiley (1917), p. 26.
193. Wiley (1930), p. 239.
194. *New York Times*, 30 December 1908; Gaughan and Barton Hutt (2004), p. 13.
195. Wiley (1930), p. 240.
196. Ibid., p. 241.
197. Gaughan and Barton Hutt (2004), p. 14.
198. Wiley (1930), pp. 262 ff.
199. Gaughan and Barton Hutt (2004), p. 14.
200. Wiley (1930), p. 311.
201. Smith (2001) pp. 85–89.

5: Mock Goslings and Pear-nanas

1. Rodden (2007), p. 31.
2. Orwell (2001), p. 190.
3. Rodden (2007), p. 31.
4. Orwell (2001), chap. 12.
5. "Germany Today Is the Land of the Ersatz," *Sheboygan Press*, 4 February 1918.
6. "Everyday Life in Berlin," *The Times*, 28 January 1918.
7. Davis (2000), pp. 89, 204.
8. Ibid., p. 205.
9. Ibid.
10. Berghoff (2001), p. 182.
11. Ibid., p. 181.
12. Spencer (2002), p. 288.
13. Beeton (2000), p. 89.
14. Humble (2005), pp. 33, 34.
15. Spencer (2002), p. 288.
16. Humble (2005), p. 95.
17. Patten (1985), pp. 16, 52.
18. Humble (2005), p. 95.

19. Clifton and Spencer (1993), p. 444.
20. Spencer (2002), p. 316.
21. Ibid., p. 319.
22. Clifton and Spencer (1993), p. 441.
23. Patten (1985), p. 9.
24. Cooper (1999), p. 127.
25. Cooper (1967), p. 92.
26. Fistere (1952), p. 166.
27. Junod (1999), p. 2.
28. "The Government Helps," *Derning Headlight*, 11 September 1931.
29. Willis (1946), p. 20.
30. Haber (2002), chap. 5.
31. Turner (1970), p. 50.
32. Kallet and Schlink (1933), p. 13 and chap. 2.
33. http://www.fda.gov/oc/history.
34. Barton Hutt (1978), p. 517.
35. Martin (1954), p. 124.
36. Faunce (1953), p. 719.
37. Garstang (1954), pp. 94–95.
38. Fistere (1952), p. 167.
39. Barton Hutt (1978), p. 510.
40. Levenstein (1993), p. 101.
41. *Long Beach Press Telegram*, 10 January 1952; *New Mexican*, 29 May 1952; *Frederick Post*, 23 February 1952; *Van Nuys News*, 17 January 1952.
42. Quoted Levenstein (1993), p. 111.
43. Levenstein (1993), p. 113.
44. Turner (1970), p. 1.
45. *Food, Drug and Cosmetics Law Journal*, 1952, p. 32.
46. Levenstein (1993), p. 109.
47. Somers (1970), p. 85.
48. Levenstein (1993), p. 113.
49. FDA website, accessed 14 October 2006.
50. Degnan (1991), p. 554.
51. *Food and Drug Law Journal*, January 1959, p. 7.
52. Turner (1970), p. 8.
53. Junod (1999), p. 8.
54. Barton Hutt (1978), p. 533.
55. Turner (1970), p. 2.
56. http://www.nns.nih.gov/1969/.
57. White House (1969), p. 120.
58. Levenstein (1993), p. 13.
59. Ibid., p. 22.

60. Whitley (2006), chap. 1.

61. Levenstein (1993), pp. 22, 23.

62. "Enriched Bread Is Great Boon to National Diet," *Clearfield Progress*, 17 April 1941.

63. Nestlé (2002), p. 302.

64. Spiekermann (2006), pp. 162–63.

65. Ibid., p. 163.

66. *Syracuse Herald Journal*, 1 August 1940, p. 29.

67. *Ogden Standard*, 22 March 1940, p. 28.

68. British Nutrition Foundation (1994), pp. 18–19.

69. Nestle (2002), p. 313.

70. http://www.newstarget.com.

71. "Vitamin Overdose." *New Scientist*, 22 April 2000.

72. "Vitamin Deaths," *New Scientist*, 24 November 2001.

73. White House (1969), pp. 118, 123.

74. Nestle (2002), p. 304.

75. http://www.bbc.co.uk/1/hi/health.

76. Nestle (2002), p. 314.

77. White House (1969), p. 123.

78. Ibid.

79. *Food and Drug Law Journal* (1970), p. 222.

80. Turner (1970), p. 5.

81. Levenstein (1993), p. 172.

82. Turner (1970), pp. 12–13.

83. "A New Sweetener, Discovered by Accident in the Lab, Goes on the Market," *New York Times*, 19 March 1975.

84. Smyth (1982–83), p. 634.

85. "FDA Banning Saccharin Use on Cancer Links," *New York Times*, 10 March 1977.

86. "Output Is Ending for Some Goods; Saccharin Substitutes to Be Used," *New York Times*, 11 March 1977.

87. "'Scientists Seek New Sweetener to Replace Saccharin," *New York Times*, 9 April 1977.

88. Smyth (1982–83), p. 636.

89. Ibid., p. 638.

90. "Sweetener Manufacturer Disputes Validity of New Health Research," *Guardian*, 30 September 2005.

91. See http://www.aspartamesafety.com.

92. *Guardian*, 30 September 2005.

93. See http://www.aspartame.info.

94. "Study Finds No Cancer Link to Sweetener," *New York Times*, 8 April 2006.

95. Butchko (1997).

96. "Sainsbury's Takes the Chemicals Out of Cola," *Daily Mail*, 23 April 2007.

97. Hesser (2003), p. 25.

98. Pyke (1970), p. 130.

99. *Flavour Industry*, vol. 4, no. 3 (March 1973), p. 119.

100. *International Flavours and Food Additives*, vol. 7, no. 4 (July/August 1976), p. 173.

101. *Flavour Industry*, vol. 4, no. 5 (May 1973), p. 214.

102. http://www.givaudan.com; accessed October 2006.

103. Corbin (1986), p. 198; Suskind (1986), p. 39.

104. Pyke (1970), p. 106.

105. *Flavour Industry*, vol. 4, no. 5 (May 1973), p. 215.

106. Ziegler and Ziegler (1998), p. 325.

107. Clarke (1922), pp. 126, 121.

108. Reproduced from Pyke (1970), p. 97.

109. Lawrence (1986), p. 38.

110. Cannon (1989), p. 116.

111. Taylor (1980), p. 36.

112. Ziegler and Ziegler (1998), p. 662.

113. Staff of the Legislation Unit (2000), p. 6.

114. *International Flavours and Food Additives*, vol. 6, no. 5 (September/October 1975), p. 297.

115. *International Flavours and Food Additives*, vol. 6, no. 4 (July/August 1975).

116. *Flavour Industry*, vol. 3, no. 10 (October 1972), p. 510.

117. Cited in Jacobsen (2005).

118. *Flavour Industry*, vol. 3, no. 10 (October 1972), p. 510.

119. *Flavour Industry*, vol. 6, no. 2 (February 1975).

120. Ziegler and Ziegler (1998), p. 369.

121. *Flavour Industry*, vol. 3, no. 10 (October 1972), p. 510.

122. Ziegler and Ziegler (1998), p. 496.

123. *Flavour Industry*, vol. 1, no. 11 (November 1970), p. 752.

124. *Flavour Industry*, vol. 6, no. 3 (March 1975).

125. Ziegler and Ziegler (1998), p. 1.

126. Cited Jacobsen (2005).

127. *Flavour Industry*, vol. 2, no. 11 (November 1971), p. 630.

128. Ziegler and Ziegler (1998), p. 211.

129. Ecott (2002), p. 211.

130. Staff of the Legislation Unit (2000), p. 16.

131. Taylor (1980), p. 44.

132. Ecott (2002), p. 212.

133. Ibid., p. 213.

134. *Flavour Industry*, vol. 3, no. 11 (November 1972), pp. 21–23.

135. Shlosser (2001), p. 127.

136. *Flavour Industry*, vol. 4, no. 8 (August 1973), pp. 334–36.

137. Nader (2000), pp. 261–65.

138. Ibid., p. 262.

139. Ibid., p. 263.

140. "Nader, the Man and the Legend," *Daily Times-News*, Burlington, 12 August 1970.

141. "Nader's Raiders," *The Times*, 6 February 1971.

142. McCarry (1972), pp. 292–93.

143. "Nader's Raiders."

144. "2 Cereal Critics to Push Efforts," *New York Times*, 7 August 1970.

145. "Breakfast Cereal Critic Cites Wide Improvement," *New York Times*, 5 November 1971.

146. "Red Dye No 2: The 20-Year Battle," *New York Times*, 28 February 1976.

147. "Red Dye 40 Called a Hazard to Health," *New York Times*, 17 December 1976.

148. Lawrence (1986), p. 59; Swanson, Kinsbourne (1980); Mandel (1994).

149. See http://teachers.net.gazette; http://www.nacsg.org.uk.

150. London Food Commission (1988), p. 53.

151. Cannon (1989), p. 24.

152. Walker in Lawrence (1986), p. 13.

153. Cannon (1989), p. 20.

154. Walker in Lawrence (1986), p. 18.

155. Cannon (1989), p. 103.

156. Cooper (1967), p. 17.

157. *Food Trade Review*, May 1975, p. 46.

158. Cannon (1989), p. 103.

159. Ibid., p. 21.

160. Walker in Lawrence (1986), p. 14.

161. Ibid., p. 20.

162. Ibid., p. 19.

163. Ibid., p. 20.

164. Cannon (1989), p. 130.

165. Ibid., p. 109.

6: Basmati Rice and Baby Milk

1. Ratledge (2004).

2. http://www.chewonthis.org.uk, accessed December 2006.

3. Shipperbottom (1993).

4. "Hazard Reported in Apple Chemicals," *New York Times*, 2 February 1989.

5. "Chilean Fruit Pulled from Shelves," *New York Times*, 15 March 1989.

6. "Cooking Oil Scandal Hurts Spain Abroad," *New York Times*, 19 October 1981.

7. BBC News, 8 September 2000.

8. BBC News, 24 February 2005.

9. BBC News, 19 July 1999.

10. Australian ABC News, 31 July 2006.

11. Coppin and High (1999), preface.

12. Kessler (1993), p. 8.

13. Nestle (2002), p. 249.

14. "FDA Finds Most Comply on Labels," *New York Times*, 14 December 1994.

15. Lyons and Rumore (1993), p. 183.

16. http://www.opsi.gov.uk/si/si1996/Uksi_19961499_en_13.htm#sdiv7, accessed January 2007.

17. Mark Woolfe, interview with author, November 2006.

18. "Police Fear 'Kebab Mafia'' behind Putrid Meat Trade," *Guardian*, 7 September 2006.

19. "Remote Control," *Guardian*, 15 May 2004.

20. Willard (2001), p. 103.

21. "Uzès Journal," *New York Times*, 6 February 2004.

22. "Tavera Journal; On Trailblazing Corsica, Sausages with Pedigree," *New York Times*, 10 July 2003.

23. "MAFF UK—Instant Coffee Surveillance Exercise," May 1994, food surveillance information sheet.

24. Mark Woolfe, interview with author, November 2006.

25. "Survey of Undeclared Horsemeat or Donkeymeat in Salami or Salami-Type Products," FSA, December 2003.

26. Li-Chan (1994).

27. Ravilious (2006), p. 2.

28. Padovan (2003); Martín (1998).

29. Woolfe and Primrose (2004), p. 222.

30. Bligh (2000), passim.

31. Barton Hutt (1978), p. 525.

32. Ibid.

33. "New York City Plans Limit on Restaurants' Use of Trans Fats," *New York Times*, 27 September 2006.

34. Nestle (2006), p. 123.

35. Reed (1975).

36. Lawrence (2004), p. 109.

37. Whitley (2006), chap. 1.

38. Ibid.

39. Ibid.

40. Davies (2004).

41. Purvis (2005).

42. Ibid.

43. Davies (2004).

44. Purvis (2005).

45. Ibid.

46. Picard (2002).

47. Purvis (2005).

48. Beaudoin (2000), p. 245.

49. Ibid.

50. Quoted in http://www.foe.co.uk/resource/press_releases/eu_commission_calls_gm_ con.html, accessed January 2007.

51. http://www. commondreams.org/headlines06/0727-06.htm, accessed January 2007.

52. Ely (1990), p. 9.

53. Barton Hutt (1972), p. 522.

54. *New York Times*, 19 February 1972.

55. "Filth," *Sunday Gazette-Mail*, West Virginia, 20 February 1972.

56. Barton Hutt (1978), p. 522.

57. http://www.cfsan.fda.gov/dms/dalbook.html#CHPTA.

58. Lyon (1994).

59. Quoted in http//www.dietdetective.com, accessed January 2007.

60. London Food Commission (1988), p. 81.

61. Ibid., p. 83.

62. http://www.soilassociation.org/web/sa/saweb.nsf/, accessed January 2007.

63. Nestle (2006), p. 465.

64. Cited in Burros (2003).

65. *Guardian*, 22 September 2006.

66. Burros (2003).

67. Cooper (2000), p. 206.

68. The term was invented by Steven Bratman, M.D.; see http://www.orthorexia.com.

69. "When Healthy Eating Turns into a Disease," *Guardian*, 10 October 2006.

70. "Original Essay on Orthorexia," http://www.orthorexia.com.

71. Nina Planck, Op-Ed, *New York Times*, 23 November 2005.

72. "Celebrity Restaurant Fined over Fake "Organic" Dishes," *The Times*, 19 December 2006.

73. http://www.comcom.govt.nz/MediaCentre/MediaReleases/200506/fakeorganicslandbutcherwithmeaty10.aspx, accessed January 2007.

74. "Millions of Faked Freerange Eggs Dupe Shoppers," *Daily Telegraph*, 17 November 2006.

75. http://www.pekingduck.org/archives/004389.php, accessed January 2007.

76. http://www.thanniennews.com, 7 April 2005, accessed January 2007.

77. E-mail to the author, 13 January 2007.

78. *The Times*, 24 July 2006.

79. http://thescotsman.scotsman.com/index.cfm?id=625522004, accessed January 2007; Reuters, 4 December 2006.

80. http://www.voanews.com/english/2007-01-10-voa16.cfm, accessed January 2007.

81. "Ex-Chief of China Food and Notes Drug Unit Sentenced to Death for Graft," *New York Times*, 30 May 2007; "Death Sentence for Drugs Chief Who Took Bribes to Clear Killer Medicines," *The Times*, 30 May 2007.

82. http://thescotsman.scotsman.com/index.cfm?id=625522004, accessed January 2007.

83. Accum, *Treatise on Adulterations* (1820), p. 31

84. "China Faces Fake Baby Milk Scandal," Food Production Daily.com, 22 April 2004.

85. "Infants in Chinese City Starve on Protein-Short Formula," *New York Times*, 5 May 2004.

86. "China 'Fake Milk' Scandal Deepens," BBC News, 22 April 2004.

87. "Infants in Chinese City Starve on Protein-Short Formula."

88. "Shopkeeper Punished for Fake Milk Powder," *China Daily*, 6 August 2004.

89. "Chinese Milk Blamed for 50 Deaths," *Guardian*, 21 April 2004.

90. Ibid.

91. "Infants in Chinese City Starve on Protein-Short Formula."

92. *New Age Metro*, 19 August 2005.

93. *Bangladesh Independent*, 3 November 2003.

94. *Daily Star*, 7 February 2006.

95. *Bangladesh Observer*, 2 July 2006.

96. *Daily Star*, 7 February 2006.

Epilogue: Adulteration in the Twenty-first Century

1. Sachs (2005).

2. Nestle (2002).

Bibliography

Accum, Friedrich Christian, *Chemical Amusement, comprising a series of curious and instructive experiments in chemistry, which are easily performed, and unattended with danger* (London: T. Boys, 1817).

———, *Culinary Chemistry: Exhibiting the Scientific Principles of Cookery* (London: R. Ackermann, 1821).

———, *An Explanatory Dictionary of the Apparatus and Instruments Employed in the Various Operations of Philosophical and Experimental Chemistry* (London: T. Boys, 1824).

———, *A Practical Treatise on Gas-Light* (London: Ackermann, 1815).

———, *A Treatise on Adulterations of Food and Culinary Poisons* (London: Longman, Hurst, Rees, Orme & Brown, 1820).

———, *A Treatise on Adulterations of Food, and Culinary Poisons*, facsimile of the first American edition, published in Philadelphia in 1820 (New York: Mallinckrodt, 1966).

———, *A Treatise on the Art of Brewing* (London: Longman, Hurst, Rees, Orme Brown, 1820).

———, *A Treatise on the Art of Making Good and Wholesome Bread of Wheat, Oats, Rye, Barley and Other Farinaceous Grain* (London: T. Boys, 1821).

———, *A Treatise on the Art of Making Wine from Native Fruits* (London: Longman, Hurst, Reese, Orme & Brown, 1820).

Acton, Eliza, *The English Bread-Book for Domestic Use Adapted to Families of Every Grade* (London: Longman, Brown, Green, 1857).

———, *Modern Cookery for Private Families*, facsimile of 1855 edition (Lewes: Southover Press, 1963).

Addison, J., and R. Steele (eds.), *The Tatler*, collated edition (London: Longman Dodsley etc., 1797).

Allingham, William, *Blackberries Picked Off Many Bushes by D. Pollex and Others, Put in a Basket by W. Allingham* (London: G. Philip & Son, 1884).

Anderson, Oscar, E., *The Health of a Nation: Harvey W. Wiley and the Fight for Pure Food* (Chicago: University of Chicago Press, 1958).

Anderson, R. C. (ed.), *The Assize of Bread Book, 1477–1517* (Southampton: Cox Sharland, 1923).

Anon. (probably Dr Peter Markham), *Poison Detected of Frightful Truths* (London: Dodsley, Osborne, Corbet, Griffith, Jones, 1757).

Anon., *Deadly Adulteration and Slow Poisoning; or Disease and Death in the Pot and the Bottle, by an enemy of fraud and villainy* (London: Sherwood, Gilbert & Piper, 1830).

Anon., *The Tricks of the Trade in the Adulterations of Food and Physic* (London: David Bogue, 1851).

Anon., *Adulteration of Food, Drink and Drugs, Being the Evidence Taken before the Parliamentary Committee* (London: David Bryce, 1855a).

Anon., *Language of the Walls: and A Voice from the Shop Windows, or The Mirror of Commercial Roguery "by one who thinks aloud"* (Manchester: Abel Heywood, 1855b).

Apple, Rima, D., *Mothers and Medicine, A Social History of Infant Feeding 1890–1950* (Madison: University of Wisconsin Press, 1987).

Atkin, Tim, and William Lee, "Just add Antifreeze," *Observer Food Monthly*, 16 January 2005.

Atkins, P. J., "Sophistication Detected or, the Adulteration of the Milk Supply, 1850–1914," *Social History*, vol. 16 (1991): 317–39.

Barton Hutt, Peter, "The Basis and Purpose of Government Regulation and Misbranding of Food," *Food Drug and Cosmetic Law Journal*, vol. 33 (1978).

Beaudoin, Kirsten, "On Tonight's Menu: Toasted Cornbread with Firefly Genes? Adapting Food Labeling Law to Consumer Protection Needs in the Biotech Century," *Marquette Law Review* (2000).

Beeton, Isabella, *Mrs Beeton's Book of Household Management*, abridged edition (London: Penguin, 2000).

Berghoff, Hartmut, "Enticement and Deprivation: The Regulation of Consumption in Pre-War Nazi Germany," in *The Politics of Consumption*, edited by Martin Daunton and Matthew Hill (Oxford: Berg, 2001), chap. 8.

Bligh, H.F.J., "Detection of Adulteration of Basmati Rice with Non-premium Long-grain Rice," *International Journal of Food Science and Technology*, vol. 35 (2000): 257–65.

Block, Daniel, "Milk," in *The Oxford Encyclopedia of Food and Drink in America*, edited by Andrew F. Smith (Oxford: Oxford University Press, 2004).

Brantz, Dorothy, "Dehumanizing the City: The Problem of Livestock in Nineteenth-Century Paris and Berlin," conference paper, European Association for Urban History, Stockholm, 2006.

British Nutrition Foundation, *Food Fortification* (London: British Nutrition Foundation, 1994).

Brosco, Jeffrey, "The Early History of the Infant Mortality Rate in America," *Pediatrics*, vol. 103 (1999): 478–85.

Browne, C. A., "The Life and Chemical Services of Frederick Accum," *Journal of Chemical Education*, vol. 2 (1925): 829–51, 1008–34, 1140–49.

Burnett, John, *Plenty and Want: A Social History of Food in England from 1815 to the Present Day*, originally published 1968 (London: Routledge, 1989).

Burros, Marion, "Is Organic Food Provably Better?" *New York Times*, 16 July 2003.

Butchko, Harriet, H., "Safety of Aspartame," letter to *Lancet*, 12 April 1997.

Camporesi, Piero, "Bread of Dreams," *History Today*, vol. 39, no. 4 (April 1989): 14–21.

Cannon, Geoffrey, *The Good Fight: The Life and Work of Caroline Walker* (London: Ebury Press, 1989).

Catchpole, Andrew, "A Vintage Year for Cheating," *Guardian*, 6 July 2006.

Cato, *Cato the Censor on Farming*, translated by Ernest Brehaut (New York: Columbia University Press, 1933).

Chevallier, Jean-Baptiste Alphonse, *Dictionnaire des altérations et falsifications des substances alimentaires, médicamenteuses et commerciales* (Paris: Béchet jeune, 1854).

Child, Samuel, *Every Man His Own Brewer*, sixth edition (London: J. Ridgeway, 1798).

Clarke, A., *Flavouring Materials: Natural and Synthetic* (London: Hodder & Stoughton, 1922).

Clayton, Edwy Godwin, *Arthur Hill Hassall: Physician and Sanitary Reformer* (London: Baillière, Tindall & Cox, 1908).

Clifton, Claire, and Colin Spencer, *The Faber Book of Food* (London: Faber & Faber, 1993).

Cole, R. J., "Frederick Accum: A Biographical Study," *Annals of Science*, vol. 7 (1951): 128–43.

———, "Sir Anthony Carlisle, FRS (1768–1840)," *Annals of Science*, vol. 8 (1952): 255–70.

Collins, Emmanuel, *Lying Detected; or some of the most frightful untruths that ever alarmed the British metropolis fairly exposed* (Bristol: E. Farley & Son, 1758).

Columella, *De re rustica, On Agriculture*, with an English translation, 3 vols. (London: Heinemann, 1954–55).

Cooper, Artemis, *Writing at the Kitchen Table: The Authorized Biography of Elizabeth David* (London: Michael Joseph, 1999).

Cooper, Derek, *The Bad Food Guide* (London: Routledge, 1967).

———, *Snail Eggs and Samphire: Dispatches from the Food Front* (London: Macmillan, 2000).

Coppin, Clayton, and Jack C. High, *The Politics of Purity: Harvey Washington Wiley and the Origins of Federal Food Policy* (Ann Arbor: University of Michigan Press, 1999).

Corbin, Alain, *The Foul and the Fragrant: Odour and the Social Imagination* (London: Berg, 1986).

David, Elizabeth, *English Bread and Yeast Cookery*, new American edition (Newton, MA: Biscuit Books, 1994).

Davies, Catriona, "Chicken Not Such a Healthy Option," *Daily Telegraph*, 4 April 2004.

Davis, Belinda, J., *Home Fires Burning: Food, Politics and Everyday Life in World War I Berlin* (Chapel Hill: University of North Carolina Press, 2000).

Degnan, Frederick H., "Rethinking the GRAS Concept," *Food and Drug Law Journal*, vol. 46 (1991): 553–82.

Denby, Daniel, "Uppie Redux? Upton Sinclair's Losses and Triumphs," *New Yorker*, 28 August 2006.

Dillon, Patrick, *Gin: The Much-Lamented Death of Madame Geneva, the Eighteenth-Century Gin Craze* (Boston: Justin Charles, 2004).

Dorey, Margaret, "Corrupt and Naughty Wares: Rhetoric versus Reality in Regulating the Food Market of Seventeenth-century London," paper given to the Oxford Symposium on Food and Cookery, September 2007.

Drummond, J. C., and Anne Wilbraham, *The Englishman's Food: A History of Five Centuries of English Diet* (London: Jonathan Cape, 1939).

Ecott, Tim, *Vanilla: Travels in Search of the Luscious Substance* (London: Michael Joseph, 2002).

Eisinger, Josef, "Lead and Wine: Eberhard Gockel and the *Colica Pictonum*," *Medical History*, vol. 26 (1982): 279–302.

Eliot, George, "Address to Working Men by Felix Holt," first published in *Blackwood's Magazine*, 1868, in George Eliot, *Essays and Leaves from a Notebook* (London: Blackwood, 1884).

Ellender, David, "A Class-Action Lawsuit against Aspartame Manufacturers: A Realistic Possibility or Just a Sweet Dream for Tort Lawyers?" *Regent University Law Review* (2005–6): 179–208.

Ely, Clansen, "Regulation of Food Additives and Contaminants in the United States," in *Symposium Proceedings: Food Law: US–EC* (Campden Food and Drink Research Association, 1990).

Engels, Friedrich, *The Condition of the Working Class in England*, edited with an introduction by David McLellan (Oxford: Oxford University Press, 1993).

Everard, Stirling, *The History of the Gas-Light and Coke Company, 1812–1949* (London: Ernest Benn, 1949).

Fabian Society, "Municipal Slaughterhouses," *Fabian Tracts*, no. 92 (1899).

Faunce, George, "The Imitation Jam Case," *The Food and Drug Cosmetic Law Journal*, vol. 8 (1953): 717–20.

Ferrières, Madeleine, *Sacred Cow, Mad Cow: A History of Food Fears*, translated by Jody Gladding (New York: Columbia University Press, 2006).

Fielden, Christopher, *Is This the Wine You Ordered, Sir?* (London: Christopher Helm, 1989).

Filby, Frederick Arthur, *A History of Food Adulteration and Analysis* (London: G. Allen & Unwin, 1934).

Fildes, Valerie, *Breasts, Bottles and Babies: A History of Infant Feeding* (Edinburgh: Edinburgh University Press, 1986).

Fistere, Charles, M., "The Imitation Jam Case—Some Implications," *The Food, Drug and Cosmetic Law Journal*, vol. 7 (1952): 165–71.

Freeman, Sarah, *Mutton and Oysters: The Victorians and Their Food* (London: Gollancz, 1989).

Galtier, C-P., *Traité de toxicologie médicale, chimique et légale et de la falsification des aliments, boissons, condiments*, 2 vols. (Paris: Chamerot, 1855).

Garnsey, Peter, *Famine and Food Supply in the Graeco-Roman World* (Cambridge: Cambridge University Press, 1988).

Garstang, Marion, R., "The Imitation Jam Decision," *The Food, Drug and Cosmetic Law Journal*, vol. 9 (1954): 92–98.

Gaughan, Anthony, and Peter Barton Hutt, "Harvey Wiley, Theodore Roosevelt and the Federal Regulation of Food and Drugs," *Food and Drug Law*, Harvard Law School (Winter 2004).

Gee, Brian, "Friedrich Christian Accum (1769–1838)," entry in the new Oxford *DNB*, 2004.

Gladwell, Malcolm, "The Ketchup Conundrum," *New Yorker*, 6 September 2004.

Gonzalez, Martín, et al., "Detection of Honey Adulteration with Beet Sugar Using Stable Isotope Methodology," *Food Chemistry*, vol. 61, no. 3 (1998): 281–86.

Goodacre, Royston, David Hammond, and Douglas Kell, "'Quantitative Analysis of the Adulteration of Orange Juice with Sucrose Using Pyrolysis Mass Spectrometry and Chemometrics," *Journal of Analytical and Applied Pyrolysis* (1997): 135–58.

Goodwin, Lorine Swainston, *The Pure Food, Drink and Drug Crusaders, 1879–1914* (London: McFarland, 1999).

Gottesman, Ronald, "Introduction" to *The Jungle* by Upton Sinclair (Harmondsworth: Penguin, 1985).

Gray, Ernest A., *By Candlelight: The Life of Arthur Hill Hassall, 1817–94* (London: Robert Hale, 1983).

Greenaway, Frank (ed.), *The Archives of the Royal Institution of Great Britain in Facsimile, 1799–1900* (Menston, Ilkley: Scolar Press, 1971).

Haber, Barbara, *From Hardtacks to Home Fries* (New York: Simon & Schuster, 2002).

Hamlin, Christopher, *A Science of Impurity: Water Analysis in Nineteenth-century Britain* (Bristol: Adam Hilger, 1990).

———, *Public Health and Social Justice in the Age of Chadwick: Britain 1800–1854* (Cambridge: Cambridge University Press, 1998).

Hassall, Arthur Hill, *Food and Its Adulterations* (London: Longman, Brown, Green, 1855).

———, *The Narrative of a Busy Life: An Autobiography* (London: Longmans Green, 1893).

Haydon, Peter, *Beer and Britannia: An Inebriated History of Britain* (London: Sutton Publishing, 2001).

Hesser, Amanda, *Cooking for Mr. Latte* (New York: Norton, 2003).

Hudson, John, *The History of Chemistry* (London: Macmillan, 1992).

Hughson, D., *The New Family Receipt Book* (London: W. Pritchard, 1817).

Humble, Nicola, *Culinary Pleasures* (London: Faber, 2005).

Hutchins, B. L., *The Public Health Agitation: 1833–1848* (London: Fifield, 1909).

Jackson, Henry, *An Essay on Bread, wherein the Bakers and Millers Are Vindicated from the Aspersions Contained in Two Pamphlets* (London: J. Wilkie, 1758).

Jacobsen, Jan Krag, paper given to the Oxford Food Symposium on Adulteration, September 2005.

Johnson, Hugh, *The Story of Wine* (London: Mitchell Beazley, 1989).

Junod, Suzanne White, "The Rise and Fall of Federal Food Standards in the United States: The Case of the Peanut Butter and Jelly Sandwich," paper given at the Spring Conference, Society for the Social History of Medicine, 9 April 1999.

Juvenal, *Juvenal: The Satires*, edited by E. Courtney (Rome, 1984).

Kallet, Arthur, and F. Klink, *100 000 000 Guinea Pigs* (New York: Vanguard Press, 1933).

Kaplan, Steven Laurence, *The Bakers of Paris and the Bread Question, 1700–1775* (Durham, NC: Duke University Press, 1996).

Kennett, Frances, *History of Perfume* (London: Harrap, 1975).

Kessler, David, "Remarks by the Commissioner of Food and Drugs," *Food and Drug Law Journal*, vol. 48 (1993): 1–11.

Keuchel, Edward, F., "Chemicals and Meat: The Embalmed Beef Scandal of the Spanish-American War," *Bulletin of Medical History*, vol. 48 (1974): 249–64.

Kolko, Gabriel, *The Triumph of Conservatism: A Re-interpretation of American History, 1900–1916* (New York: Free Press of Glencoe, 1963).

Lawrence, Felicity (ed.), *Additives: Your Completes Survival Guide* (London: Century, 1986).

———, *Not on the Label: What Really Goes into the Food on Your Plate* (London: Penguin, 2004).

Lee, Kwang-Sun, "Infant Mortality Decline in the Late Nineteenth and Early Twentieth Century: Role of Market Milk," http://www.ironwood.cpe.uchicago.edu/CPE_Workshop/paper, accessed 9 October 2006.

Letheby, Henry, *On Food: Four Cantor Lectures* (London: Longman, Green, 1870).

Levenson, Barry, M., *Habeas Codfish: Reflections on Food and the Law* (Madison: University of Wisconsin Press, 2001).

Levenstein, Harvey, *Paradox of Plenty: A Social History of Eating in Modern America* (Oxford: Oxford University Press, 1993).

Li-Chan, Eunice, "Developments in Detection of Adulteration of Olive Oil," *Trends in Food Science and Technology*, vol. 5 (January 1994): 3–11.

London Food Commission, *Food Adulteration and How to Beat It* (London: Unwin Hyman, 1988).

Loomis, C. Grant, "Mary Had a Parody: A Rhyme of Childhood in Folk Tradition," *Western Folklore*, vol. 17 (January 1985): 45–51.

Loubère, Leo A., *The Red and the White: A History of Wine in France and Italy in the Nineteenth Century* (Albany: State University of New York Press, 1978).

Lyon, William, F., "Insects as Human Food," Ohio State University Fact Sheet, 1994, http://www.ohioline.osu.edu, accessed November 2006.

Lyons, Jean. and Martha Rumore, "Food Labeling Then and Now," *Journal of Pharmacy and Law*, vol. 172 (1993).

McCance, R. A., and E. M. Widdowson, *Breads White and Brown: Their Place in Thought and Social History* (London: Pitman Medical Publishing, 1956).

McCarry, C., *Citizen Nader* (New York: Saturday Review Press, 1972).

McGee, Harold, *On Food and Cooking: The Science and Lore of the Kitchen* (New York: Simon & Schuster, 1984).

MacKenney, Richard, *Tradesmen and Traders: The World of Guilds in Venice and Europe c. 1280–c. 1650* (London: Croom Helm, 1987).

Mandel, Boris L., "Food Additives Are Common Causes of Attention Deficit Disorder in Children," *Annals of Allergy* (May 1994).

Manning, James, *The Nature of Bread Honestly and Dishonestly Made* (London: R. Davis, 1757).

Markham, Peter, *A Letter to the Right Honourable William Pitt, Relating to the Abuses Practiced by Bakers* (London: M. Cooper, 1757).

Martin, John, B., "The Imitation Jam Case and Its Effect on the Ice Cream Industry," *Food and Drug Law Journal*, vol. 8 (1954): 123–26.

Mayhew, Henry, *The Morning Chronicle Survey of Labour and the Poor: The Metropolitan Districts* (London: Caliban Books, 1980–).

Miller, Ian, "Alum Production at Carlton Alum Works," http://www.oxfordarch .co.uk (article of May 2004).

Mitchell, John, *Treatise on the Falsifications of Food and the Chemical Means Employed to Detect Them* (London: Hippolyte Baillière, 1848).

Monckton, H. A., *A History of English Ale and Beer* (London: Bodley Head, 1966).

Murphy, Kevin C., "Pure Food, the Press and the Poison Squad: Evaluating Coverage of Harvey W. Wiley's Hygienic Table," 2001," http://www.kevincmurphy.com, accessed September 2006.

Nader, Ralph, *The Ralph Nader Reader* (London: Seven Stories Press, 2000).

Nelson, Robert, L., "The Price of Bread: Poverty, Purchasing Power and the Victorian Labourer's Standard of Living," modified 25 December 2005, http://www .victorianweb.org.

Nestle, Marion, *Food Politics* (Berkeley: University of California Press, 2002).

———, *What to Eat* (New York: North Point Press, 2006).

Nightingale, Pamela, *A Medieval Mercantile Community: The Grocer's Company and the Politics and Trade of London 1000–1485* (New Haven: Yale University Press, 1995).

Normandy, Alphonse, *The Commercial Handbook of Chemical Analysis* (London: George Knight & Sons, 1850).

357

Olver, Lynne, "Mock Foods," in *The Oxford Encyclopedia of Food and Drink in America* (Oxford: Oxford University Press, 2004).

Orwell, George, *The Road to Wigan Pier*, originally published 1937 (Harmondsworth: Penguin, 2001).

Padovan, G. J., et al., "Detection of Adulteration of Commercial Honey Samples by the 13C/12C Isotopic Ratio," *Food Chemistry*, vol. 82, no. 4 (2003): 633–36.

Parmentier, Antoine-Augustin, "Treatise on the Composition and Use of Chocolate," *Nicholson's Journal: Journal of Natural Philosophy, Chemistry and the Arts*, vol. 5 (1803).

Patten, Marguerite, *We'll Eat Again* (London: Hamlyn, 1985).

Patton, Jeffrey, *Additives, Adulterants and Contaminants in Beer* (London: Patton Publications, 1989).

Phillips, Rod, "Wine and Adulteration," *History Today* (June 2000): 31–37.

———, *A Short History of Wine* (London: Penguin, 2001).

Picard, André, "Today's Fruits, Vegetables Lack Yesterday's Nutrition," *Toronto Globe and Mail*, 6 July 2002.

Platt, Hugh, *Sundrie New and Artificiall Remedies against Famine* (London: P.S., 1596).

Pliny, the Elder, *Naturalis historia*, Natural History, English and Latin, translated by H. Rackham, 10 vols. (London: Heinemann, 1968).

Potter, Stephen, *The Magic Number: The Story of "57"* (London: M. Reinhardt, 1959).

Purvis, Andrew, "It's Supposed to Be Lean Cuisine. So Why Is This Chicken Fatter than It Looks?" *Observer Food Monthly*, 15 May 2005.

Pyke, Magnus, *Synthetic Food* (London: John Murray, 1970).

———, *Technological Eating* (London: John Murray, 1972).

Ratledge, Andrew, "Food Fraud and the British Consumer, 1800–1860," Third International Conference of the Research Centre for the History of Food and Drink, 12–14 July 2004.

Ravilious, Kate, "Buyer Beware: The Rise of Food Fraud," *New Scientist*, 15 November 2006.

Redding, Cyrus, *History and Description of Modern Wines* (London: Whittaker, Treacher and Arnot, 1833).

Reed, Gerald (ed.), *Enzymes in Food Processing* (New York: Academic Press, 1975).

Renard, Georges, *Guilds in the Middle Ages*, translated by Dorothy Terry, originally published 1918 (New York: Augustus M. Kelly, 1968).

Reynolds, John Hamilton, *The Press, or Literary Chit-Chat. A Satire* (London: Lupton Relfe 1822).

Roberts, H. J., "Aspartame and Brain Cancer," letter to *Lancet*, 1 February 1997.

Robinson, Jancis, *The Oxford Companion to Wine*, second edition (Oxford: Oxford University Press, 1999).

Rodden, John (ed.), *The Cambridge Companion to George Orwell* (Cambridge: Cambridge University Press, 2007).

Roosevelt, Theodore, *Theodore Roosevelt: An Autobiography* (London: Macmillan, 1913).

———, *The Letters of Theodore Roosevelt*, edited by Elting E. Morison (Cambridge: Harvard University Press, 1954).

Rothschild, Louis, "The Newest Regulatory Agency in Washington," *Food and Drug Law Journal*, vol. 33 (1978): 86–93.

Rowlinson, P. J., "Food Adulteration: Its Control in Nineteenth Century Britain," *Interdisciplinary Science Reviews*, vol. 7, no. 1 (1982): 63–71.

Rubin, Miri, *The Hollow Crown: A History of Britain in the Late Middle Ages* (London: Allen Lane, 2005).

Rumohr, Carl Friedrich von, *The Essence of Cookery*, translated by Barbara Yeomans (Totnes: Prospect Books, 1993 [originally published Stuttgart, 1822]).

Rundell, Mrs., *A New System of Domestic Cookery* (London: John Murray, 1818).

Sabine, R. H., "The Changing Role of the Flavourist," *Flavour Industry*, vol. 3, no. 10 (October 1972): 509–10.

Sachs, Jeffrey, *The End of Poverty: Economic Possibilities for Our Time* (London: Allen Lane, 2005).

Schlosser, Eric, *Fast Food Nation: What the All-American Meal Is Doing to the World* (London: Allen Lane, 2001).

Schmid, Ron, "Pasteurize or Certify: Two Solutions to the Milk Problem," http://www.realmilk.com/untoldstory, accessed 9 October 2006.

Shaugnessy, Haydn, "What's Your Poison?" *Irish Times*, 18 October 2005.

Shipperbottom, Roy, "Paradise Lost: The Adulteration of Spices," *Oxford Symposium on Food and Cookery* (1993), pp. 247–53.

Sinclair, Upton, *The Jungle* (London: T. Werner Laurie, 1906).

———, *The Autobiography of Upton Sinclair* (London: W. H. Allen, 1963).

Singer, Peter, and Jim Mason, *Eating: Why We Eat What We Eat and Why It Matters* (London: Arrow Books, 2006).

Slotnick, Bonnie J. "Sylvester Graham," in *The Oxford Encyclopedia of Food and Drink in America*, edited by Andrew F. Smith (Oxford: Oxford University Press, 2004), vol. 1, pp. 573–74.

Smith, Andrew, *Pure Ketchup: A History of America's National Condiment* (Washington, DC: Smithsonian Institution Press, 2001).

Smith, R.E.F., and David Christian, *Bread and Salt: A Social and Economic History of Food and Drink in Russia* (Cambridge: Cambridge University Press, 1984).

Smollett, Tobias, *The Expedition of Humphry Clinker* (London: W. Johnston, 1771).

Smyth, Todd R., "The FDA's Public Board of Inquiry and the Aspartame Decision," *Indiana Law Journal*, vol. 58 (1982–83): 627–49.

Soffritti, Morando, et al., "First Experimental Demonstration of the Multipotential Carcinogenic Effects of Aspartame Administered in the Feed to Spague- Dowley Rats," *Environmental Health Perspectives* (November 2005): 1–34.

Somers, Ira A., "Additives, Standards and Nutritional Contributions of Foods," *Food Drug and Cosmetic Law Journal* (1970).

Spencer, Colin, *British Food: An Extraordinary Thousand Years of History* (London: Grub Street, 2002).

Spiekermann, Uwe, "Brown Bread for Victory: German and British Wholemeal Politics in the Inter-War Period," in *Food and Conflict in Europe in the Age of Two World Wars*, edited by Frank Trentmann and Flemming Just (London: Palgrave, 2006).

Staff of the Legislation Unit, *Flavourings in Food—A Legal Perspective* (Leatherhead, Survey: Leatherhead Publishing, 2000).

Stanziani, Alessandro (ed.), *La Qualité des produits en France (XVIIIe–XXe siècles)* (Paris: Belin, 2003).

——, *Histoire de la qualité alimentaire, XIXe–XXe siècles* (Paris: Seuil, 2005).

Stieb, Ernst Walter, *Drug Adulteration: Detection and Control in Nineteenth-Century Britain* (Madison: University of Wisconsin Press, 1966).

Studer, P., *The Oak Book of Southampton*, vol. 2, *A Fourteenth-Century Version of the Medieval Sea-Laws Known as the Rolls of Oleron* (Southampton: Cox & Sharland, 1911).

Suh, Suk Bong, *Upton Sinclair and The Jungle* (Seoul: American Studies Institute, 1997).

Sullivan, Mark, *Our Times: The United States, 1900–1925*, vol. 2 (New York: C. Scribner's Sons, 1927).

Suskind, Patrick, *Perfume: The Story of a Murderer*, translated by John E. Woods (London: Penguin, 1986).

Swanson, Heather, *Medieval Artisans: An Urban Class in Late Medieval England* (Oxford: Basil Blackwell, 1989).

Swanson, J., and M. Kinsbourne, "'Food Dyes Impair Performance of Hyperactive Children on a Laboratory Learning Test," *Science* (March 1980): 1485–87.

Taylor, Arthur, J., *Laissez-faire and State Intervention in Nineteenth-century Britain* (London: Macmillan, 1972).

Taylor, R. J., *Food Additives* (Chichester: John Wiley & Sons, 1980).

Tickletooth, Tabitha, *The Dinner Question or How to Dine Well & Economically*, facsimile edition (Blackawton, Devon: Prospect Books, 1999).

Toussaint-Samat, Maguelonne, *A History of food*, translated by Anthea Bell (Oxford: Blackwell, 1992).

Turner, James, S., *The Chemical Feast: The Ralph Nader Study Group Report on Food Protection and the Food and Drug Administration* (New York: Grossman, 1970).

Twain, Mark, *Life on the Mississippi*, facsimile of original 1883 edition (Oxford: Oxford University Press, 1996).

Waldron, H. A., "James Hardy and the Devonshire Colic," *Medical History* (1969): 74–81.

White House, *The White House Conference Report on Nutrition and Health: Full Report*, "The New Foods Panel" (1969), pp. 116ff (available at http://www.nns.nih .gov/1969/full_report/PDFcontents.htm, accessed June 2007).

Whitley, Andrew, *Bread Matters: The State of Modern Bread and a Definitive Guide to Baking Your Own* (London: Fourth Estate, 2006).

Whittet, T. D., "Pepperers, Spicers and Grocers—Forerunners of the Apothecaries," *Proceedings of the Royal Society of Medicine*, vol. 61 (August 1968): 801–6.

Wiley, Harvey W., *Foods and Their Adulteration* (London: J. & A. Churchill, 1907).

———, *1001 Tests of Foods, Beverages and Toilet Accessories* (New York: Hearst's International Library, 1914).

———, *Foods and Their Adulteration*, third edition (London: J. & A. Churchill, 1917).

———, *Beverages and Their Adulteration* (London: J. & A. Churchill, 1919).

———, *An Autobiography* (Indianapolis: Bobbs-Merrill, 1930).

Willard, Pat, *Saffron: The Vagabond Life of the World's Most Seductive Spice* (London: Souvenir Press, 2001).

Willis, Daniel P., "Preventing Economic Adulteration of Food," *Food and Drug Law Journal* (March 1946).

Woolfe, Mark, and Sandy Primrose, "Food Forensics: Using DNA Technology to Combat Misdescription and Fraud," *Trends in Biotechnology*, vol. 22, no. 5 (May 2004): 222–26.

Young, James Harvey, *Pure Food: Securing the Pure Food and Drugs Act of 1906* (Princeton: Princeton University Press, 1989).

Zehetner, Anthony, and Mark McLean, "Aspartame and the InterNet," letter to *Lancet*, 3 July 1999.

Ziegler, Erich, and Herta Ziegler (eds.), *Flavourings: Production, Consumption, Applications, Regulations* (Chichester: Wiley-VCH, 1998).

Zupko, Ronald Edward, *British Weights and Measures: A History from Antiquity to the Seventeenth Century* (Madison: University of Wisconsin Press, 1977).

Acknowledgments

I am grateful to my publishers, John Murray in the UK and Princeton University Press in the United States. Anya Serota thought of the title and supported the project in its early stages. At Princeton University Press, many thanks to Ian Malcolm for generously taking the book on and to Ellen Foos and Anita O'Brien for their hard work and professionalism. I am extremely fortunate in my agents: in the UK, Pat Kavanagh, who has a better sense of food than anyone I know; and in the United States, Emma Parry, whose ability to order well in restaurants is just one of her peerless qualities.

Many friends, relatives, and colleagues helped supply details, advice, and support of various kinds. Thanks to Anna Ashelford at the Food Standards Agency, Adam Balic (for fake eggs), Lucy Bannell, Catherine Blyth, Caroline and Hugh Boileau, Jenny Brannan at the British Medical Association, Geoff Brennan (for norms), Bronwen Bromberger, Lauren Carleton Paget, Emily Charkin, Hilary Cox, Caroline Davidson, Charlotte Dewar, Margaret Dorey, Lindsay Duguid (for Karl Marx), Katherine Duncan-Jones, Nathalie Golden at the Food Standards Agency, Theo Fairley, Judith Flanders (for red anchovies), Bob Goodin (for philosophy), Christelle Guibert at Waitrose, Barry Higman, Sarah Howard (for wine), Tristram Hunt (for Engels), Mark Lake (for dyed olives), Miranda Landgraf, Melissa Lane, Dan Lepard (for bread), Paul Levy, Esther McNeill, Anne Malcolm (for mock turtle soup), Anthea Morrison, Anna Murphy, Francis Percival, Ruth Platt, Elfreda Pownall, Rose Prince (for enzymes), Kelly Rayney at Waitrose, Claudia Roden (for saffron), Emma Rothschild, Tim Rowse, Miri Rubin, Garry and Ruth Runciman, Magnus Ryan, Abby Scott (for hot chocolate), Ruth Scurr, Marlena Spieler, Gareth Stedman Jones (for guilds),

Peter Stodhart (for Juvenal), Adam Tooze, Frank Trentmann, Mark Turner, Simon and Toby Welfare, Andrew Whitley (for fortification), Andrew Wilson, Emily Wilson, Stephen Wilson.

I would also like to acknowledge the help of a number of individuals and organizations who supplied me with information or assistance, including Colman's Mustard, Ossau Iraty Cheese, Tilda Rice, and Waitrose. Mark Leatham of Merchant Gourmet was generous with his time and knowledge. I am very grateful to the Food Standards Agency in Britain and to the Food and Drug Administration in the United States (particularly Cindy E. Lachin) for supplying me with various pieces of information. At the Food Standards Agency, I am particularly indebted to Mark Woolfe. My thanks to the research school of history at the Australian National University in Canberra, Australia, where I wrote one of the chapters. Staff at the Cambridge University Library were exceptionally helpful.

My greatest thanks must go to David, Tom, and Natasha Runciman, who eat my food every day and haven't died yet.

Picture Credits

The author and publishers would like to thank the following for permission to reproduce illustrations: pp. 6, 8, 13, 22, 37, 108, 140 and 230, the Syndics of the Cambridge University Library; p. 12, © Museum of London; pp. 61 and 312, Waitrose; p. 97, Mansell Collection/ Time & Life/Getty Images; p. 98 © ARPL/TopFoto; p. 138, Colman's; pp. 148, 167, 173 and 207, The Advertising Archives; pp. 175, 183, 184, 201, 275 and 277, U.S. Food and Drug Administration; pp. 192 and 194, Upton Sinclair Collection, The Lilly Library, Indiana University, Bloomington, Indiana; p. 215, akg-images/ullstein bild; p. 244, PEPSI, PEPSI-COLA, DIET PEPSI and the PEPSI Globe design are registered trademarks of PepsiCo, Inc. Used with permission; pp. 249 and 255, the Syndics of the Cambridge University Library and Sensient Flavors Ltd; p. 284, Istara, Ossau-Iraty/BMA Communications; p. 293, Tilda; p. 294, Food Standards Agency; p. 315, CHINA OUT REUTERS/ China Newsphoto; p. 317, REUTERS/China Photos ASW/TW. The images on pp. 32, 144–45 and 219 are from the author's collection.

Every effort has been made to clear permissions. If permission has not been granted please contact the publisher who will include a credit in subsequent editions.

Index

abattoirs (municipal slaughterhouses), in Paris, 200

Accum, Christian (Herz Marcus), 10

Accum, Friedrich Christian (Frederick): adulteration of food, popular impact of book on, 1–6; adulteration of food, the chemistry of, 16–20; on adulteration of food in Britain, 20, 25–33; battle against food swindlers, impact of his disgrace on, 42–45; on beer, 19, 39; on bread adulterated with alum, 83; career of, 6–15; *Chemical Amusement,* 14; as chemist, 2, 7, 9, 11–15, 117–18; on coffee, 23–24; as consultant, 17–18, 40; on cooking and food in Britain, 21–25; *Culinary Chemistry,* 21–22, 43; demon grocers, confirmation of suspicions regarding, 99; disgrace and departure from England of, 39–45; *Explanatory Dictionary of the Apparatus and Instruments Employed . . . by a Practical Chemist,* 13; flavour adjusters, exposure of, 248; food, love of, 1–2, 10–11, 21, 23, 45; gas lighting, promotion of, 7–9; government, call for action from, 34–35, 95; knowledge, fighting adulteration with, 325; lemonless lemonade, complaint about, 233; portraits of, 6, 12; *Practical Treatise on Gas-Light,* 8–9; swindlers, motivation of, 316; *System of Theoretical and Practical Chemistry,* 15; tea and sloe leaves, distinguishing between, 32–33; *A Treatise on Adulterations of Food, and Culinary Poisons,* 1–6, 16–20, 25, 27–28, 42–43, 55, 116; Walker and, 268, 270; Wiley, comparison to, 174; wine,

adulteration of, 55, 58; wine, on using fruits to make, 23, 59

Accum, Judith (née Judith Suzanne Marthe Bert la Motte), 10–11

Accum, Phillip, 10

acetimeter, Chevalier's, 118

Ackermann, Rudolph, 7, 9, 41

Acton, Eliza, 21, 83, 107–8, 110, 112

Adams, Congressman from New York, 154

Adams, Samuel Hopkins, 189

Addison, Joseph, 55–56

additives. *See* synthetic food

adulteration: Accum on (*see* Accum, Friedrich Christian (Frederick)); aesthetic, U.S. regulations regarding, 305–7; in Asia, 313–21; authenticity as the opposite of, 286 (*see also* Food Standards Agency (FSA)); battle against, government and (*see* government; names of countries); battle against, impact of Accum's disgrace on, 42–45; of bread during times of famine, 74–76; in Britain (*see* Britain); chemical additives (*see* synthetic food); the chemistry of, 15–20; commercial defense of, xiv, 19–20, 33–35, 38–39, 95–96, 141; common familiarity with, xi, 4–5; definitions of, 137, 139, 141; ersatz food as, 214–18, 247–48, 314 (*see also* synthetic food); genetic modification as, 304–5; greed as the motive for, xii–xiii, 4, 27, 85–86, 164, 316, 322 (*see also* swindling and swindlers); grocers and, 96–101; intent as an element in, xii; of make-believe, mock food as, 219–20; pesticides as constituting and a remedy for, 305 (*see also*